Climate Change Impacts and Women's Livelihood

Very few studies have been conducted to explore the vulnerability of women in the context of climate change. This book addresses this absence by investigating the structure of women's livelihoods and coping capacity in a disaster vulnerable coastal area of Bangladesh.

The research findings suggest that the distribution of livelihood capitals of vulnerable women in rural Bangladesh is heavily influenced by several climatic events, such as cyclones, floods and seasonal droughts that periodically affect the region. Women face several challenges in their livelihoods, including vulnerability to their income, household assets, lives and health, food security, education, water sources, sanitation and transportation systems, because of ongoing climate change impacts. The findings have important policy relevance for all involved in disaster and risk management, both within Bangladesh and the developing countries facing climate change impacts.

Based on the research findings, the book also provides recommendations to improving the livelihoods of women in the coastal communities. This book will appeal to academics, researchers and professionals in environmental management, gender and development, and climate change governance looking at the effects of and adaptation to climate change, gender issues and natural disaster management strategies.

Salim Momtaz is an Associate Professor in Sustainable Resource Management at the School of Environmental and Life Sciences, University of Newcastle, Australia.

Muhammad Asaduzzaman is a Conjoint Lecturer at the School of Environmental and Life Sciences, University of Newcastle, Australia.

Routledge Studies in Hazards, Disaster Risk and Climate Change

Series Editor: Ilan Kelman

Reader in Risk, Resilience and Global Health at the Institute for Risk and Disaster Reduction (IRDR) and the Institute for Global Health (IGH), University College London (UCL)

This series provides a forum for original and vibrant research. It offers contributions from each of these communities as well as innovative titles that examine the links between hazards, disasters and climate change, to bring these schools of thought closer together. This series promotes interdisciplinary scholarly work that is empirically and theoretically informed, with titles reflecting the wealth of research being undertaken in these diverse and exciting fields.

For more information about this series, please visit:
www.routledge.com/Routledge-Studies-in-Hazards-Disaster-Risk-and-Climate-Change/book-series/HDC

Climate Change Impacts and Women's Livelihood

Vulnerability in Developing Countries

**Salim Momtaz and
Muhammad Asaduzzaman**

LONDON AND NEW YORK

First published 2019
by Routledge
2 Park Square, Milton Park, Abingdon, Oxon OX14 4RN

and by Routledge
52 Vanderbilt Avenue, New York, NY 10017

First issued in paperback 2020

Routledge is an imprint of the Taylor & Francis Group, an informa business

British Library Cataloguing-in-Publication Data
A catalogue record for this book is available from the British Library

Library of Congress Cataloging-in-Publication Data
A catalog record for this book has been requested

ISBN 13: 978-0-367-58469-6 (pbk)
ISBN 13: 978-1-138-61610-3 (hbk)

Typeset in Times New Roman
by Apex CoVantage, LLC

Contents

Figures

Tables

Photos

Preface

Bangladesh is frequently cited as a country that is most vulnerable to climate change. In Bangladesh, most of the adverse effects of climate change occur in the form of extreme weather events, such as cyclone, flood, drought, salinity ingress, riverbank erosion and tidal surge, leading to large-scale damage to crops, employment, livelihoods and the national wellbeing. Although it is generally stated that women are relatively more vulnerable than men in the context of climate change, few studies have been conducted to closely examine this statement. The present study investigates the structure of women's livelihoods, livelihood vulnerabilities and coping capacity in the context of climate variability and change in a disaster vulnerable coastal area of Bangladesh. The use of the concepts of the Sustainable Livelihood Framework (SLF) and Disaster Crunch Model (DCM) allowed for a greater understanding of these issues on the ground. Two livelihood vulnerability indexes (LVIs), namely the LVI and the Intergovernmental Panel on Climate Change – Livelihood Vulnerability Index (IPCC-LVI), are used to measure the degree of livelihood vulnerability of women in the study area. The results show that the distribution of five livelihood capitals (human, natural, financial, social and physical) of women are heavily influenced by several climatic events, such as cyclones that periodically affect the region. Women also face several vulnerabilities in their livelihoods, including vulnerability to their income, household assets, lives and health, food security, education, water sources, sanitation and transportation systems, because of ongoing climate change impacts. They only have limited adaptation strategies that enable them to reduce the climate-related risks. However, they do practise some traditional coping strategies to respond to the increasing effects of climate change. While quantifying the degree of vulnerability, both of the indexes indicate a 'high vulnerability' level in regards to women's livelihoods. In particular, women are more vulnerable in terms of physical and financial capitals in their present livelihood system. The results indicate that it is of paramount importance to instigate strategies to help build the adaptive capacity of women to reduce the burden created by their livelihood vulnerability. Overall, this book contributes empirical evidence to current debates in the literature on climate change by enhancing an understanding of the characteristics and determinants of livelihood vulnerability of women in the coastal areas of Bangladesh. The findings have important policy relevance for all involved in disaster and risk management, both within Bangladesh and the developing countries facing climate change impacts. The findings of this book also allow identification of a range of measures that could be utilised to help address the impacts of current and future climate variability and change in regards to women's livelihoods, particularly in the poorer, rurally based coastal communities. Based on the research findings, the book also provides some recommendations to improving the livelihoods of women in the coastal communities.

Acknowledgements

During the preparation of this book and the several fieldwork and visits to various locations for data collection we received support and guidance from many people and organisations. We are grateful to the people who participated in this research project, especially the women in the study sites in Bangladesh, for their hospitality and cooperation towards us.

Acronyms and abbreviations

ADB	Asian Development Bank
BBS	Bangladesh Bureau of Statistics
BCAS	Bangladesh Centre for Advanced Studies
BCCSAP	Bangladesh's Climate Change Strategy and Action Plan
BDT	Bangladeshi Taka
BGMEA	Bangladesh Garment Manufacturers and Exporters Association
CCC	Climate Change Cell
CPD	Centre for Policy Dialogue
DAW	United Nations Division for Advancement of Women
DCM	Disaster Crunch Model
DFID	Department for International Development
FAO	Food and Agricultural Organisation
FGD	Focus Group Discussion
GAR	Global Assessment Report
GDP	Gross Domestic Product
GoB	Government of Bangladesh
HH	Household
IDS	Institute of Development Studies
IFAD	International Fund for Agricultural Development
IPCC	Intergovernmental Panel on Climate Change
IUCN	International Union for Conservation of Nature
KII	Key Informant Interview
LVI	Livelihood Vulnerability Index
MJF	Manusher Jonno Foundation
MoDMR	Ministry of Disaster Management and Relief
MoEF	Ministry of Environment and Forestry
MoF	Ministry of Finance
MoFDM	Ministry of Food and Disaster Management
MPCS	Multi-purpose Cyclone Centre
NAPA	National Adaptation Plan of Action
NGF	Nowabenki Gonomukhi Foundation
NGO	Non-Government Organisation
NIPORT	National Institute of Population Research and Training

NTFPs	Non-Timber Forest Products
OECD	Organisation for Economic Cooperation and Development
PRDI	Participatory Research and Development Initiative
RMG	readymade garments
SLF	Sustainable Livelihood Framework
SRDI	Soil Research Development Institute
UN	United Nations
UNDP	United Nations Development Programme
UNEP	United Nations Environment Programme
UNESCO	United Nations Educational, Scientific and Cultural Organization
UNFCCC	United Nations Framework Convention on Climate Change
UNICEF	United Nations Children's Fund
UNIFEM	United Nations Development Fund for Women
UNISDR	United Nations International Strategy for Disaster Reduction
USAID	United States Agency for International Development
VGD	Vulnerable Group Development
WEDO	Women's Environment and Development Organization
WHO	World Health Organization

1 Introduction

1.1 Background to the study

Bangladesh is frequently cited as one of the most vulnerable countries to climate change, due to its disadvantageous geographic location: flat and low-lying topography; high population density; high levels of poverty; reliance of many livelihoods on climate-sensitive sectors, particularly agriculture and fisheries; and inefficient institutional aspects (Climate Change Cell, 2006). Many of the anticipated adverse effects from climate change, such as salinity intrusion due to sea level rise, higher temperatures, enhanced monsoon precipitation and an increase in cyclone intensity, will aggravate the existing stresses that already impede development in Bangladesh (Ministry of Environment and Forest, 2005). Ultimately, these impacts could be extremely detrimental to the economy, the environment, national development and the people of Bangladesh now and into the future (Reid & Sims, 2007).

The Intergovernmental Panel on Climate Change (IPCC) has predicted that Bangladesh is slated to lose large amounts of its land due to rising sea levels and, if sea levels rise up to one metre this century, Bangladesh could lose up to 20 per cent of its land mass and up to 30 million Bangladeshis could become climate refugees (Harasawa, 2006). Cyclonic winds are also likely to increase in intensity, because of the positive correlation with sea surface temperature (United Nations Development Programme, 2007). Sea level rise is a critical issue for large population in coastal areas and islands particularly. Inhabitants living on low-lying coastal plains are also at risk from floods and displacement from the coastal zone. Rising temperatures in the region are also likely to continue with global warming. The effect on future rainfall is uncertain, but future climate change could have a profound impact on the monsoon, which underpins the rainfall regime. In addition, droughts, cyclones and intense rainfall events, saltwater intrusion and erosion are also likely to continue to increase (IPCC, 2001; IPCC, 2007).

In Bangladesh, 140,000 people died from the flood-related effects of Cyclone Gorky in 1991. Within that number, women outnumbered men by approximately 93 per cent (14:1 in ratio). Women also represented an estimated 61 per cent of fatalities in Myanmar after Cyclone Nargis in 2008 and 70 per cent of those who died during the 2004 Indian Ocean tsunami in Banda Aceh, Indonesia (World

Bank, 2011a). Reid and Sims (2007) suggest that in 2007, during Cyclone Sidr and the subsequent floods in Bangladesh, the death rate was reportedly five times higher for women than for men. This was because women were not allowed to leave their homes in the absence of a male relative and so many waited (until it was too late) for their male relatives to return. Another report says the effects of longer-term climate change may be felt more acutely by women where their endowments, agency and opportunities are not equal to or are more climate-sensitive than those of men (World Bank, 2011b). Changes to water resources and hydrology will also have a major impact on Bangladesh, where people depend on the surface water for fish cultivation, navigation, industry and other uses, and where the groundwater is used for domestic purposes and irrigation. Women are the main users and carriers of water. As the availability and quality of water declines and resources become scarcer, women will suffer increasing workloads to collect unsalinated water to sustain their families (Huq & Ayers, 2008). Overall, women also have less exposure to disaster-related training and information, such as early warning systems, than men (UN, 2009).

A 2007 study of 141 natural disasters over the years 1981–2002 found that, when economic and social rights are realised equally for both sexes, disaster-related death rates do not differ significantly for men or women. However, when women's rights and socio-economic status are not equal to those of men, more women die in disasters (Neumayer & Plümper, 2007). In the context of climate change, the World Bank makes three major statements: (i) women are disproportionately vulnerable to the effects of natural disasters and climate change where their rights and socio-economic status are not equal to those of men; (ii) empowerment of women is an important ingredient in building climate resilience; and (iii) low-emissions development pathways can be more effective and more equitable where they are designed using a gender-informed approach (World Bank, 2011b). Evidence is mounting that empowering women to create institutional platforms that expand their own, their families' and their communities' endowments, agency and opportunities can serve as a powerful springboard for building climate resilience more generally (World Bank, 2011b). Therefore, there is an urgent need to understand women's livelihoods and vulnerability to grow awareness and address their vulnerabilities appropriately, and build resilience to climate change for the sustainable future of Bangladesh.

1.2 Rationale for this book

The impact of climate change is extremely visible and already felt in Bangladesh. Available literature suggests that women in particular are the most vulnerable group to be affected by climate change impacts. Women are affected differently and more severely than men by climate change and natural disasters due to: high gender-based discrimination against women; unequal power relations between men and women; less access to assets and resources; and, overall, fewer capabilities and opportunities for adjustment to related vulnerabilities (UN, 2009).

With 160 million people, Bangladesh is the world's densest nation, with the exception of the city states (e.g. Singapore, Hong Kong). Moreover, about one-quarter of the country's Gross Domestic Product (GDP) comes from agriculture, which makes the country's economy relatively sensitive to climate variability and change (World Bank, 2011c). The per capita income in Bangladesh is US$ 1110 (2013–14) and around one-third (32 per cent) of the people in Bangladesh live below the poverty line (Bangladesh Bureau of Statistics, 2015). Gender-poverty links show that around 70 per cent of the poor in the world are women and that their vulnerability is accentuated by some factors such as race, ethnicity and age (CCC, 2009). In Bangladesh, women constitute almost half of the total population and they are not only socially discriminated against, but are also subject to a variety of threats, exploitation and harassment. In particular, rural women are educationally, politically and socially disadvantaged, resulting in economic dependency and increased vulnerabilities (Sarker, 2007). The status of Bangladeshi women has been ranked as the lowest in the world on the basis of 20 indicators related to health, marriage, children, education, employment and social equality (UN Women, 2000). Growing empirical evidence supports the broad view that women's overall lower access to assets and services, and their levels of political and social recognition make them more vulnerable than men to the effects of climate change and natural disasters (World Bank, 2011b).

Although it is generally stated that women are relatively more vulnerable than men in the context of climate change, few studies so far have examined this statement, especially in Bangladesh. Until recently, studies on climate change have mostly focused on the physical consequences of climate change and their technical solutions, rather than on gender dimensions. Women in particular suffer differently from the impacts of climate change; they have different needs, knowledge, management practices and strategies for coping with the manifestations of climate change. There has, however, been a very limited amount of research in Bangladesh that has analysed the positive and negative impacts of climate change on gender. A gender-disaggregated approach is thus required to shed more light on the livelihood of women and the degree of livelihood vulnerability that exist, as well as to identify appropriate adaptation mechanisms.

The aim of this book is to analyse the impacts of climate change on the livelihoods of women in the disaster vulnerable coastal areas of Bangladesh. It investigates the structure of women's livelihoods, vulnerabilities and their adaptive capacity to climate change by applying a gender-sensitive Sustainable Livelihood Framework (SLF). The study will empirically measure the outcome of different livelihood strategies of women to climate change in a specific context, that is, in disaster vulnerable coastal areas of Bangladesh. The findings will help to address climate change problems more specifically, raise awareness and improve knowledge in the context of gender sensitivity. It will also provide policy makers with a future action plan which considers adaptive capacity and opportunities to reduce climate-induced risks for women throughout the nation.

1.3 Layout of the book

This book is divided into eight chapters. A brief description of each chapter is provided here.

Chapter 2 provides a brief overview of Bangladesh to introduce the country to the reader. The chapter then focuses on the key areas of the climate change effects in Bangladesh that have been experienced in the last decades. A review on the status of the rural women in Bangladesh is presented in the next section. The chapter then briefly discusses the climate change effects on women's livelihoods overall. In Chapter 3 a discussion on the theoretical and methodological aspects is included to provide the global context of the themes discussed in this book. Under the theoretical framework, an understanding of the SLF that is used to assess the livelihood of women in the context of climate variability and extremes is provided. This chapter then briefly discusses the concept of vulnerability in the context of climate change, and presents a Disaster Crunch Model (DCM) that helps to assess vulnerability. In addition, the chapter explains two methods of vulnerability measurements, the Livelihood Vulnerability Index (LVI) and the IPCC-LVI.

Chapter 4 presents an overview of the geographical and socio-economic setting of the study area, and how the livelihoods of people in this rural area are connected to serious climate change effects. Chapter 5 investigates the livelihoods of women in the study area following SLF approach. Five livelihood capitals – namely human, natural, financial, social and physical – that directly help to construct livelihood, are assessed to understand women's particular livelihoods in a disaster susceptible setting. Moreover, the chapter compares the livelihoods of women in relation to a trigger point (Cyclone Aila in 2009) to see how a climate-induced disaster like this may impact upon and change the livelihood setting of women in the study area. Chapter 6 identifies the key vulnerabilities of women's livelihoods in the context of climate change. It explains how different climate change events affect the livelihoods of women and their coping mechanism. The chapter also examines women's positions in respect to their access to major welfare facilities that can potentially help reduce their climate-induced vulnerabilities. Chapter 7 presents a concrete measurement of vulnerability of women by computing an LVI and an IPCC-LVI. These two quantitative measures give vulnerability scores based on some components and subcomponents of livelihood capitals by which the degree of vulnerability can be realised. Chapter 8 draws conclusions on the livelihood vulnerability of women in a disaster-prone coastal area of Bangladesh that can be generalised for similar geographical and cultural settings in developing areas. Recommendations are made for improving the livelihood of rural women in the coastal areas of Bangladesh.

References

BBS. (2015). *Highlights*. Dhaka: Bangladesh Bureau of Statistics, Statistics and Informatics Division (SID), Ministry of Planning, Government of the People's Republic of Bangladesh. Retrieved from www.bbs.gov.bd/home.aspx

Climate Change Cell (CCC). (2006). *Who is doing what in Bangladesh?* Dhaka: Comprehensive Disaster Management Programme, Ministry of Environment and Forests, Government of the People's Republic of Bangladesh.

Climate Change Cell (CCC). (2009). *Climate change, gender and vulnerable groups in Bangladesh*. Dhaka: Department of Environment, Ministry of Environment and Forests, Government of the People's Republic of Bangladesh.

Harasawa, H. (2006). Key vulnerabilities and critical levels of impacts on East and Southeast Asia. In H. J. Schellnhuber, W. Cramer, N. Nakicenovic, T. Wigley, & G. Yohe (Eds.), *Avoiding dangerous climate change* (pp. 243–249). Cambridge and New York: Cambridge University Press.

Huq, S., & Ayers, J. (2008). *Climate change impacts and responses in Bangladesh*. Briefing note prepared for the European Parliament, DG Internal Policies of the Union, Policy Department, Economic and Scientific Policy, European Parliament.

IPCC. (2001). *Climate change 2001: Impacts, adaptation and vulnerability*. Contribution of Working Group II to the fourth assessment report of the Intergovernmental Panel on Climate Change (IPCC) (1032 pp.). Cambridge and New York: Cambridge University Press.

IPCC. (2007). *Climate change 2007: The physical science basis*. Contribution of Working Group I to the fourth assessment report of the Intergovernmental Panel on Climate Change (IPCC) (996 pp.). Cambridge and New York: Cambridge University Press.

MoEF. (2005). *National adaptation programme of action (NAPA): Final report*. Dhaka: Ministry of Environment and Forests (MoEF), Government of the People's Republic of Bangladesh.

Reid, H., & Sims, A. (2007). *Up in smoke? Asia and the Pacific: The threat from climate change to human development and the environment*. The Fifth Report from the Working Group on Climate Change and Development, Intergovernmental Panel on Climate Change (IPCC) (92 pp.).

Sarker, S. (2007, July 17–20). *Globalization and women at work: A feminist discourse*. Paper presented at International Feminist Summit, Townsville, Australia.

UN (2009). Towards a Gender-Sensitive Agenda for Energy, Environment and Climate Change. United Nations Expert Paper. p. 3.

UN Women. (2000, June 5–9). *The 23rd special session of the UN general assembly with the theme "Beijing + 5 – women 2000: Gender equality, development and peace for the twenty-first century,"* held in the General Assembly, New York.

UNDP. (2007). *Human development report 2007–08: Fighting climate change: Human solidarity in a divided world*. New York: United Nations Development Programme, Palgrave Macmillan.

World Bank. (2011a). *Making women's voices count: Integrating gender issues in disaster risk management in Asia and Pacific region* (Operational Guidance Notes). Washington, DC: The World Bank.

World Bank. (2011b). *Gender and climate change: Three things you should know* (p. 7). Washington, DC: The Social Development Department, The World Bank.

World Bank. (2011c). *World development report 2012*. Washington, DC: Gender Equality and Development, The World Bank.

2 Climate change impacts and women in Bangladesh

This chapter provides an overview of Bangladesh and its related climate change effects. In addition, it discusses the status of rural women in Bangladesh who reside within the context of vulnerability and risk created by climate change, with a particular focus on the specific climate change impacts on them.

2.1 Background information on Bangladesh

2.1.1 Geographical location of Bangladesh

Bangladesh is a South Asian country located between 20 and 26 degrees north latitude and 88 and 92 degrees east longitude. It is formed by a deltaic plain at the confluence of the Ganges (Padma), Brahmaputra (Jamuna) and Meghna Rivers and their tributaries (Global Assessment Report, 2011). The total land area, which is 147,570 square kilometres, consists mostly of floodplains (Figure 2.1).

Bangladesh has a 580-km coastal line in the Bay of Bengal of the Indian Ocean. The coastal belt comprises 30 per cent of its geographical area and is home to one-third of the country's population (BBS, 2013a). Mean elevations range from less than one metre on tidal floodplains, one to three metres on the main river and estuarine floodplains, and up to six metres in the Sylhet basin in the north-east (Rashid, 1991). About 10 per cent of the country is about one metre above the mean sea level (MSL), and one-third is under tidal excursions. Overall, it is this kind of geographic setting and location that makes Bangladesh more vulnerable than many other countries to climate change, especially for water-related hazards (Global Assessment Report, 2011).

2.1.2 Weather and seasons in Bangladesh

Bangladesh experiences moderate temperature increases in post-monsoon seasons and strong warming (0.1°C–0.3°C per decade) during monsoon seasons. There has been a general rising trend in surface temperature in the order of 0.5°C ±0.1°C over the country and the entire South Asian region during the past century (IPCC, 2007b). The available literature suggests an increase in temperature of 0.5°C to 2.0°C by 2030, and a rise in sea level of between 30 and 150 cm by 2050 (IPCC,

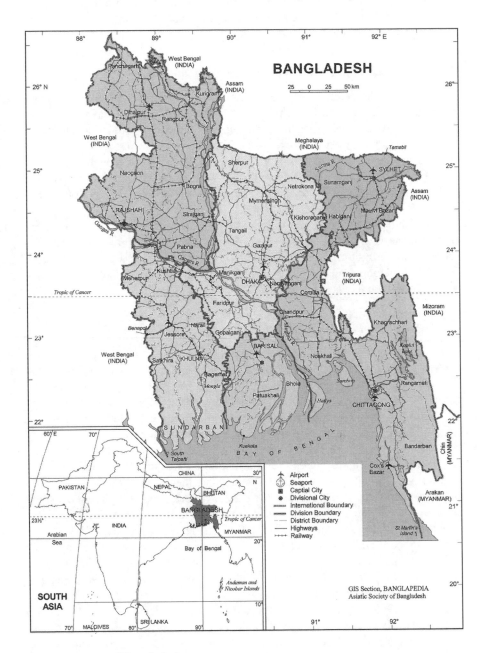

Figure 2.1 Map of Bangladesh

(Source: Banglapedia, 2015)

2007a). Most climate modelling predicts an increase in average rainfall from 8 per cent to 15 per cent by 2030. Increased rainfall is already resulting in more extreme flooding within the country and throughout the region.

2.1.3 Demography and overpopulation in Bangladesh

Bangladesh is the eighth most populous country in the world with a population of approximately 160 million. This huge population resides in an area of 147,570 square kilometres, making it the most densely populated country in the world (1218 persons/km^2) (World Bank, 2015). Life expectancy is around 63 years and there is an adult literacy rate of 47.5 per cent. The population growth rate is 1.47 per cent, and male-female ratio is 100:100.3 (BBS, 2015). The recent Human Development Report of United Nations ranks Bangladesh number 142 out of 185 nations (UNDP, 2014).

The Gender Inequality Index (GII) is a composite index that reflects women's disadvantage in three dimensions: reproductive health, empowerment and the labour market. In terms of the GII in 2014, Bangladesh ranked 115 in the world, whereas its neighbouring countries, Nepal, Myanmar and India, are ranked 94, 84 and 127, respectively (UNDP, 2014).

2.1.4 Poverty and inequality in Bangladesh

Bangladesh is considered as a developing country with a GDP per capita of US$1314 in 2014–15 (BBS, 2015). The percentage of people who live below the national poverty line is 31.5 per cent while 43.3 per cent of people live on less than $1.25 ppp[1] a day. While overall improvement in wellbeing can be seen across all regions, poverty continues to be a daunting problem with about 47 million people still living in poverty and 26 million people in extreme poverty.

The poverty rate is highest in rural areas, at 36 per cent, compared with 28 per cent in urban centres. Half of all rural children are chronically malnourished and 14 per cent suffer from acute malnutrition (International Fund for Agricultural Development, 2015). While some rural areas achieved growth based on new income sources, such as non-farm self-employment income, salaried wages and remittances, and money from abroad in the 1990s due to an influx of NGOs, the extremely poor could not benefit from these activities, as they had no capital or ability to take advantage of the opportunities presented. Therefore, inequalities on the whole tended to increase in rural areas throughout the 1990s. Female-headed households, illiterate people, agricultural waged labourers and the people who live in remote areas with unfavourable agricultural environments have been more vulnerable to poverty due to limited access to assets, transport, power and other infrastructures (UNDP, 2013). Poor countries and poor communities are more vulnerable to climate change impacts; thus Bangladesh's poverty situation is another reason why this country is one of the most climate vulnerable countries in the world. Poverty is especially persistent in three areas: the north-west, which is affected by droughts and river erosion; the central northern region, which is subject to serious seasonal flooding that limits crop production; and the southern coastal zones, which are affected by soil salinity and cyclones (IFAD, 2015), and comprise the study area of the present research.

2.2 Climate change and Bangladesh

The impact of climate change is frequently cited as one of the most severe environmental problems facing humanity (UNDP, 2002). The foremost evidence for worldwide climate change has been global warming. Global warming has increased and will continue to raise sea levels due to thermal expansion of the oceans and the melting of ice stored in glaciers or ice sheets. Global sea levels will rise by at least 18 cm, but in the worst case scenario as much as 59 cm by the year 2100 (IPCC, 2007a). The consequences of sea level rise include more frequent and devastating flooding, cyclones, storm surges and even complete submersion of many of the world's populous low-lying areas, such as Bangladesh and the Nile Delta region. Currently, more than 200 million people live on coastal floodplains around the world, with two million square kilometres of land and $1 trillion worth of assets being elevated at less than one metre above current sea level (Stern, 2006).

Ericson, Vorosmarty, Dingman, Ward, and Meybeck (2006) estimate that nearly 300 million people inhabit a sample of 40 deltas globally, including all the large mega deltas with an average population density of 500 people/km^2. The largest population resides in the Ganges-Brahmaputra delta (Bangladesh), where the population density is 1218 people/km^2 (World Bank, 2015). The people in this densely populated territory are severely affected by climate change variables and natural calamities. In terms of the impacts on human communities, the poor are the most vulnerable to climate change risks and impacts, as they are highly exposed to the environment through their living conditions and livelihoods. They have the fewest resources to prepare and plan for the impacts, and the lowest capacity to respond (Asian Cities Climate Change Resilience Network [ACC-CRN], 2012).

Bangladesh as a low-lying deltaic country located in the northern Indian Ocean, is ranked as the most vulnerable country to tropical cyclone risk (Peduzzi, Chatenou, Dao, De Bono, Herold, Kossin, Mouton, & Nordbeck, 2012). There is a consensus among scientists that, in South Asia, Bangladesh in particular is the most impacted by climate change amongst the Asian regions. The Fourth Assessment Report of the IPCC (2007b) states the following as the main climate change impacts in the region: increased frequency of droughts and floods affecting local production negatively; sea level rise exposing coasts to increasing risks, including coastal erosion and increasing human induced pressures on coastal areas; and glacier melt in the Himalayas, increasing flooding and rock avalanches, and decreasing crop yields by up to 30 per cent by the mid-twenty-first century. The report also maintains that within South Asia, Bangladesh is the most vulnerable country because of its regional connectivity through geo-physical and hydrological features, and its reliance on nature for livelihood.

A number of major studies in the past have investigated the causes of the vulnerability of Bangladesh due to climate change (see: Asian Development Bank, 1994; Karim, Hussain, & Ahmed, 1998; Warrick & Ahmad, 1996). Most of the adverse effects of climate change occur in the form of extreme weather events, while water-related hazards, such as floods, droughts, salinity ingress, riverbank erosion and tidal bores, are likely to be exacerbated, leading to large-scale damage to crops, employment, livelihoods and the national economy (Asaduzzaman, Reazuddin, & Ahmed,1997; Choudhury, Neelormi, Quadir, Mallick, & Ahmed, 2005). Small changes in average climatic conditions, however, can have a big influence on extremes such as floods, cyclones, water surges and droughts, and these are already noticeable in Bangladesh. These observed extreme events are briefly discussed below.

2.2.1 Cyclones and storm surges

Cyclones and storm surges threaten coastal communities worldwide. Globally, the number of cyclones has increased more than threefold from 1970 to 2009 (EM-DAT: The International Disaster Database, 2015). The strength and number of major cyclones may be increasing because of higher sea surface temperatures associated with global warming (Romm, 2012). Tropical cyclones and storm surges are particularly severe in the Bay of Bengal region. Bangladesh is especially vulnerable to cyclones because of its location at the triangular shaped head of the Bay of Bengal, the sea level geography of its coastal area, its high population density and the lack of coastal protection systems. During the pre-monsoon (April–May) or post-monsoon (October–November) seasons, cyclones frequently hit the coastal regions of Bangladesh. About 40 per cent of the total global storm surges are recorded in Bangladesh and the deadliest cyclones in the past 50 years, in terms of deaths and casualties, are those that have struck Bangladesh (Haque, Hashizume, Kolivras, Overgaard, Das, & Yamamoto, 2012).

Over five million Bangladeshis live in areas highly vulnerable to cyclones and storm surges. Roughly 55 per cent of the coastal population lives within 100 kilometres of the 710 kilometre-long coastal belt of Bangladesh. The majority of the people in this area are low-income agricultural workers and 70 per cent of them are landless and relatively asset-poor (WEDO, 2008). Bangladesh's coast has witnessed 14 serious cyclones in the last 50 years and, of these, three (Bhola in 1970, Gorky in 1991 and Sidr in 2007) were catastrophic (Khan, 2008). Cyclone Sidr (November 2007) and Cyclone Aila (May 2009) provide recent examples of devastating storm surges in Bangladesh. It has been reported by the government of Bangladesh (GoB, 2008a) that Cyclone Sidr in 2007 hit concurrently with a heavy storm of about 215 km/h, resulting in a six-metre tidal wave which affected about 6,851,147 people from 1,611,139 families in 200 upazilas/sub-districts. This cyclone resulted in 4234 casualties and 55,282 injuries (EM-DAT: The International Disaster Database, 2015). It also damaged approximately 186,971 hectares of crops and 365,670 houses (Islam, 2008). The livelihoods of 8.9 million people were affected and the damages and losses from Cyclone Sidr totalled US$

1.67 billion (GoB, 2008a). Moreover, the Sundarbans ecosystems were severely affected (up to 30 per cent damage) by Cyclone Sidr (Khatun & Islam, 2010). Later, severe category 1 Cyclone Aila hit 14 districts on the south-west coast of Bangladesh on the 25th May 2009. It was the second major blow to the region in less than two years: many of these areas were still recovering from the effects of 2007's Cyclone Sidr when Aila struck (UNICEF, 2015).

Cyclone Aila produced a wind speed of 120km/h and a storm surge three metres above the normal astronomical tide, which caused 190 deaths, 7103 injuries and affected 3.9 million people (EM-DAT: The International Disaster Database, 2015). However, during the past 20 years overall, Bangladesh has managed to reduce deaths and injuries from cyclone events. For example, the most recent severe cyclone of 2007 caused 4234 deaths, a 100-fold reduction compared with the devastating 1970 cyclone where the death toll was estimated at more than 300,000 people. In addition to the immediate death and suffering caused by such disasters, cyclones also have direct and indirect impacts on general public health, livelihoods, infrastructure, the economy and socio-cultural foundations. They can also affect access to food and drinking water, and increase the transmission risks of infectious diseases, such as diarrhoea, hepatitis, malaria, dengue, pneumonia, eye infections and skin diseases, thus contributing to the interruption of livelihoods (Haque et al., 2012).

By most estimates, the intensity and frequency of cyclones is likely to increase in Bangladesh. Mirza and Paul (1992) estimate that the frequency of cyclones rose from 0.51/year in 1877–1964 to 1.12/year during 1965–1980. The IPCC (2007b) also projected that intense and more frequent tropical cyclone activities with extreme high sea level (excluding tsunamis) are going to increase in this area. A study by BCAS-RA-Approtech (1994) also reports a 10 per cent increase in the intensity of cyclones in Bangladesh due to future climate change, leading to greater economic loss. Using dynamic Regional Climate Model (RCM)-driven simulations of current and future climates, a study by Unnikrishnan, Kumar, Fernandes, Michael, and Patwardhan (2006) indicates a significant increase in the frequency of the highest storm surges for the Bay of Bengal, despite no substantial change in the frequency of cyclones.

2.2.2 Sea level rise

It is predicted that Bangladesh will be among the most affected countries in South Asia, with an expected 2°C rise in the world's average temperatures within the next decades (World Bank, 2013). Rising sea levels, more extreme heat and intense cyclones will threaten food production, livelihoods and infrastructure, as well as slow the reduction of poverty (World Bank, 2013). It has also been estimated that, sea level rise (SLR) will lead to a potential loss of 15,668 km^2 of land, which is expected to affect 11 per cent of the population or 5.5 million people. If the sea level should rise by one metre, this will result in a 20 per cent loss of land, affecting 14.8 million people. The direct and indirect consequences of SLR include: saltwater intrusion into surface and groundwater systems, drainage congestion, and increased water logging and devastating effects on mangroves (WEDO, 2008).

A World Bank study (2000) shows that a 10-cm, 25-cm and one-metre rise in sea level by 2020, 2050 and 2100 will affect 2 per cent, 4 per cent and 17.5 per cent of total land mass respectively. The UNEP (1989) predicted that a 1.5-metre sea level rise for the Bangladesh coast by 2030 will affect 22,000 square kilometres (16 per cent of the total land mass area), with a population of 17 million (15 per cent of the total population) being affected. Since this scenario was calculated in 1989, the expected rate of sea level rise has been modified because of uncertainty. At present, the expected rates of change will most likely occur in about 150 years from now (Sarwar, 2005). However, the number of people likely to be affected according to the projection of the World Bank (2000) with a one-metre sea level rise (17.5 million) and that of the UNEP (1989) with a 1.5-metre sea level rise (17 million) is similar in expected outcomes and potential losses.

2.2.3 Floods and flooding

Bangladesh constitutes only about 7 per cent of the area of the combined catchments of three major eastern Himalayan rivers: the Ganges, the Brahmaputra and the Meghna (GBM), and the country drains over 92 per cent of the total annual flow of this GBM system. When monsoon-driven excessive runoff in these rivers combines with local rainfall, the country is frequently faced with over-bank spillages and floods, particularly along the major rivers. The current trend of flooding in Bangladesh is changing: frequency, length and intensity of floods is increasing, with more damage to people, homes, crops and other assets; thus floods have become more unpredictable in terms of onset and scale (Alam, Rahman, Farok, Kabir, &. Fatema, 2007).

The IPCC Special Report on Extreme Events (SREX), prepared by Handmer et al. (2012) estimates that in a normal year, river spills and drainage congestion caused inundation of 20–25 per cent of the country's total area, but in 1987, 1988 and 1998, floods inundated more than 60 per cent of the country causing the death toll to rise into the thousands, leaving millions homeless. The 1998 flood resulted in 1100 deaths and 30 million people were left homeless. The flooding pattern in Bangladesh points towards an increase in frequency over the years. Historical and recent data show that during the past 40 years at least, seven major floods have taken place in Bangladesh. Some of the worst ones occurred during the years of 1987, 1988, 1998, 2004 and 2007, making an average of one in six years (Ghatak, Kamal, & Mishra, 2012). Recent studies show that due to the effect of sea level rise, the densely populated coastal zone of Bangladesh is becoming highly vulnerable to coastal floods; whereas glacier melts in the Himalayan region caused flash floods in the mountainous regions and the foothills in Nepal, and this extends to the northern region of Bangladesh (Ghatak et al., 2012).

Research findings by Alam, Nishat and Siddiqui (1998) suggest that in Bangladesh, under climate change scenarios, about 18 per cent of current low-lying flooded areas will be susceptible to higher levels of flooding, while about 12 to 16 per cent of new areas which are not at risk of flooding currently, will be at risk from varied degrees of flooding in the near future. In an average hydrological year, flood-prone areas will increase from about 25 per cent to 39 per cent (Alam et al., 1998). In Bangladesh, 2007 was a unique year of disaster history. Two major extreme flood events occurred in July and August, quickly followed by the category 4 Cyclone Sidr in November. The floods alone caused 3,363 casualties, affected 10 million people and reduced crop output by at least 13 per cent (WEDO, 2008). It is reported that by 2030 an additional 14.3 per cent of the country will become extremely vulnerable to floods, while existing flood-prone areas will face increased flooding (Deb, Khaled, Al Amin, & Nabi, 2009). The coastal zone, particularly in the southern areas of the country, will also face increased water logging due to increased flood volumes draining and causing increased sea levels downstream. Thus, the probability of potential damage due to floods in Bangladesh is likely to increase as a consequence of the increase in the intensity of extreme precipitation events (Baida & Shrestha, 2007).

2.2.4 Salinity

Salinity is also becoming a long-term problem in Bangladesh and is expected to be further exacerbated by climate change and sea level rise. Salinity intrusion due to the reduction of freshwater flow from upstream, salinisation of groundwater and fluctuation of soil salinity are major concerns in Bangladesh. The anticipated sea level rise will produce salinity impacts on three fronts: surface water, groundwater and soil. Increased soil salinity due to climate change will significantly reduce food grain production (PRDI, 2015a).

Under the present climate change variability, salinity ingress has become a major hydro-geophysical and social problem in the south-western region of Bangladesh. During the dry season, salinity is more intense and a lack of suitable drinking water becomes an acute problem for affected communities. Salinity along the Bangladesh coast has already encroached over 100 kilometres inland into domestic ponds, groundwater supplies and agricultural land, through the various estuaries and water inlets intertwined with major rivers. The resultant sea-water intrusion is increasing salinity in coastal drinking water, bringing with it severe health consequences to surrounding population (Khan & Islam, 2011).

Being an agrarian nation, most of the country's land areas are used for agricultural purposes, particularly for rice production since rice is the staple food of 160 million people in Bangladesh. In the last 30 years, salinity intrusion has degraded soil quality to the extent that some farmers cannot grow any agricultural crops at all (Rashid, Hoque, & Iftekhar, 2004). In 1973, about 1.5 million hectares of land was under mild salinity; it increased to 2.5 million hectares in 1997, and to about 3.0 million in 2007. Soil salinity, water logging and acidification affect annually 3.05 million, 0.7 million and 0.6 million hectares of cropland respectively. Soil salinity will affect 13 districts in total and will cover 20.37 per cent of the total paddy area, with a potential loss of 395,000 tonnes of rice (MoEF, 2008).

Coastal people rely heavily on tube-wells, rivers and ponds for their drinking water and cooking. Although the Food and Agriculture Organisation's (FAO) allowable water salinity level for human consumption is less than half a gram per kilogram of water (parts per thousand or ppt), river salinity in some coastal districts reaches as high as four grams in the rainy season and an alarming 13 grams per kilogram in the dry season. Seawater, which is extremely harmful for humans, contains 35 grams of salinity per kilogram of water. Approximately 20 million of the 37 million people living on these coasts (over 57 per cent) are adversely affected by such salinity in their drinking water (Khan & Islam, 2011). The Sundarbans ecosystem is also at risk due to increased salinity, which in turn affects forest biodiversity and the population depending on its resources (PRDI, 2015b).

2.2.5 Drought

Causes of drought in Bangladesh are also related to climate variability and non-availability of surface water resources. The immediate cause of a rainfall shortage may be due to one or more factors, including absence of moisture in the atmosphere or large-scale downward movement of air within the atmosphere, which suppresses rainfall. Changes in such factors involve changes in local, regional and global weather and climate (FAO, 2007).

Overall, climate change may be seen to pose a serious threat to agriculture, food security and the nutritional status of ordinary people in Bangladesh. The impacts of climate variability and change in the form of drought cause additional risks for agriculture. During the last 50 years, Bangladesh has suffered about 20 periods of drought. The drought conditions in north-western Bangladesh in recent decades led to a shortfall in rice production of 3.5 million tons in the 1990s. Current severe drought can affect the yield in 30 per cent of the country, reducing national production by 10 per cent by 2030 (Bangladesh Centre for Advanced Studies, 2010). Drought affects annually about 2.32 million hectares in the *Kharif* season (November–June) and 1.2 million hectares of cropland during the *Rabi* season (July–October). About 49 districts of Bangladesh and 59 per cent of the total area used for rice production will be affected due to droughts by 2030 (MoEF, 2008; Deb et al., 2009). High temperatures followed by drought were found to reduce the yields of the high-yielding Aus, Aman and Boro rice. The estimates show that about 55 to 62 per cent of rice yields (58.5 per cent) and 2.43 per cent of total production will be affected due to drought by 2030 (MoEF, 2008; Deb et al., 2009).

2.3 The status of women in rural Bangladesh

Bangladesh is a resource-limited and overpopulated country where society is highly stratified and services and opportunities are determined by gender, class and location. However, women make up nearly half of the population, which means they present a huge potential resource to be utilised for the socio-economic development of the country. About 80 per cent of women live in rural areas (BBS, 2013c). The traditional society of Bangladesh is ruled by patriarchal values and the expected norms of female members of the society are subordination, subservience, subjugation and segregation. These result in the discrimination of women from birth, leading to the deprivation of their access to all opportunities and benefits in family and societal life; thus women are put in a most disadvantageous position (Islam & Sultana, 2006). Bangladesh's socio-cultural environment contains pervasive gender discrimination, so girls and women face many obstacles to their development.

Education is a social phenomenon that strongly influences women's control of their own future, but low levels of female education have been frequently cited in Bangladesh (Islam & Sultana, 2006). Lack of education is one of the main factors that deters women from equal participation in socio-economic activities with their male counterparts and helps to perpetuate the inequality between the sexes. The female literacy rate is lower compared to males for the age group of 15–45 years, both in rural and urban areas. The female literacy rate is 44.5 per cent in rural areas compared to 51.2 per cent for males. In urban areas, the female literacy rate is 60.8 per cent compared to 67.7 per cent for males, so a rural/urban divide is particularly obvious here (BBS, 2013b).

Bangladesh's rates of child marriage and adolescent motherhood are among the highest in the world. The legal age of marriage is 18 for girls; however, three-quarters of women aged 20–49 are married before the age of 18. The practice of arranging child marriages remains common, especially in rural areas and in urban slums, where many families believe that the onset of puberty signifies readiness for marriage (UNICEF, 2010). Bangladesh also has one of the world's highest rates of adolescent motherhood. One in four women starts childbearing before the age of 20 (NIPORT, 2015). Maternal mortality rates also remain extremely high. Poor maternal health is the result of early marriage, and women's malnutrition, a lack of access to and use of medical services and a lack of knowledge and information contributes to this (UNICEF, 2010). Most women give birth without a skilled attendant. Despite an increase in health facilities nationally, 85 per cent of deliveries still take place at home. A woman's lifetime risk of dying in pregnancy or childbirth is one in 51, compared to one in 47,600 in Ireland (the best performer) About 12,000 women die every year from pregnancy-related complications or in childbirth in Bangladesh (GoB, 2008b). Thus, overall, maternal health and mortality can be linked with women's low status in households and their restricted mobility more generally (UNICEF, 2010).

Over the years, government, NGOs and development partners have been working towards greater women's development and the implementing of a series of development interventions for eliminating gender discrimination. Secondary school education is becoming more frequent for women and, as a result, their health status is improving (ADB, 2001). A profound change for young and adolescent girls has been the introduction of cash for school and stipend programmes in the education sector, which has increased girls' attendance at school overall. Girls now outnumber boys in primary and secondary schooling (GoB, 2008b). However, net attendance rates in secondary education are still extremely low, at only 53 per cent for girls and 46 per cent for boys. In tertiary education, there are only six girls for every 10 boys, well below the Millennium Development Goal target of full equality (UNICEF and BBS, 2010). Female adult literacy rates increased from 27.4 per cent in 1997 to 53.7 per cent in 2011 (BBS, 2013b).

Increased literacy directly relates to increased employment opportunities for women. Women migrants, mostly from female-headed households (FHHs), now contribute a major share of the informal urban labour market (Gibson, Mahmud, Toufique, &Turton, 2004). Due to increased access to services and cash, more women are now able to use health services. As a result, female life expectancy increased from 58.1 years in 1997 to 70.2 years in 2011 (WHO, 2011). Women's participation in politics and administration, which was negligible in the past, also increased.

The status of a woman comes from her family and while her role includes the maintenance of her family as a social institution and as an economic entity, the decision-making powers and economic control are almost always in the hands of men. Within the family, women's roles in decision making are very low. All decisions regarding income-related activities are made by men. Often, women are not able to make any decisions about family matters without male involvement. In the past, even reproductive decisions were made by men (UNICEF, 2001). Due to dependency on a male person in their family, women's organisational participation is also low. Sen (2000) states that present gender systems are oppressive to women in two ways: i) unequal access to resources and divisions of labour within and outside the home; and ii) non-recognition of the work of household reproduction.

However, gradual change in social attitudes has occurred over the last few years. As a consequence, many women can now take advantage of new economic and social opportunities, adding significant improvements in key development indicators. Women have made important gains in the formal labour market in the past 20 years, mainly due to increased participation in the garment sector and an NGO-led microcredit revolution that targets women. The participation of women in the RMG (readymade garments) sector is well known. The RMG sector employs 3.6 million people, of which 80 per cent are women, the majority coming from rural areas (BGMEA, 2015). Gibson et al. (2004) found that key features of women's livelihoods in Bangladesh are breaking new ground; the situation for women is very dynamic as increases in non-farm work and urbanisation of rural life are affecting society more generally. Women's contribution to readymade garment exports and crop production, as well as their contribution to the remittance economy is increasing day by day. However, poor women especially, work in employment that is poorly paid, insecure and often seasonal.

Increased access to microfinance also has helped transform some women's household labour into cash contributions to the household income. Following the footsteps of the Grameen Bank's minimalist credit strategy,[2] a number of NGOs in Bangladesh (e.g. BRAC, ASA) have been targeting rural women who suffer different types of socio-economic subjugations. Although NGOs are doing a good job, it would perhaps be too much to expect that the NGOs themselves could make all rural women resourceful and empowered (Bayes, 2015). The debate tends to persist as to whether provisions of credit for poor women could ever significantly change the social equations at the village level in which this sub-set of the population live. Critics tend to suggest that while a marginal increase in income and assets can enhance wellbeing and economic security, the increase could be too little to affect the pervasively entrenched political and economic relations present there, as well as the long-held cultural beliefs (Bayes, 2015).

In spite of some positive outcomes, however, gender inequalities, especially in rural areas with respect to: enjoyment of human rights; political freedom and economic status; land ownership; housing conditions; exposure to violence; and education and health are still major concerns for Bangladesh, as overall they make women more vulnerable to social, environmental and political changes. Women's household roles (overburdened by family activities), poverty, social position and established religious norms and values act as a hindrance to overcoming these entrenched positions (Islam, 2009).

Although some rural women do have access to property, employment and credit through the country's legal system, this access remains limited by social norms and customs. Women play an important role in a wide range of income-generating activities, but their contribution to the national economy is largely unaccounted for. Apart from household work (also not recognised officially), rural women are actively involved in farm activities ranging from the selection of seed to the harvesting and storing of crops. Despite their crucial role in agriculture, the general undervaluing of women by society deprives them of equitable economic opportunities, status and access to resources. They are not even considered as farmers, as it is the men who market the produce and control the income (*The Daily Star*, 6 May 2015). In 2014, a joint research project by the Manusher Jonno Foundation (MJF) and the Centre for Policy Dialogue (CPD) revealed that women who work as unpaid workers accounted for 45.6 per cent of the total employed in agriculture. The research also estimated the value of women's unpaid work contribution to agriculture, other informal sectors and to the household to be 76.8 per cent of the total GDP for the 2013–14 fiscal year. Thus, women's contribution to these sectors must be quantified when determining a country's GDP, particularly in Bangladesh, where 89 per cent of the country's women are engaged in such informal sectors and do not get any payment for their work, despite its essential contribution to society (CPD and MJF, 2014).

Women's involvement in rural production activities includes: raising seedlings, gathering seeds, post-harvesting, cow fattening and milking, goat farming, backyard poultry rearing, pisciculture, agriculture, horticulture, food processing, cane and bam works, silk reeling, handloom weaving, garment making, fishnet making, coir production and handicrafts (WEDO, 2008). Similar activities by rural women were found in a study by Asaduzzaman, Rahman, and Jahan (2004). A significant number of rural women, particularly from extremely poor landless households, also engage in paid labour in construction and earthwork and field-based agricultural work, activities that traditionally have fallen within the male domain (WEDO, 2008). Rural women are often dependent on the natural environment for their livelihood. Maintenance of households and women's livelihoods are, therefore, directly impacted by climate-related damage to or scarcity of natural resources (UNIFEM, 2008).

Overall, however, there is not enough research into the specificities of women's livelihoods and their vulnerabilities in rural Bangladesh, especially in the coastal areas to gain an accurate picture of the situation many find themselves in. Thus, a gendered approach to research is therefore important in order to identify women's livelihood options and outcomes, vulnerabilities to crisis, and their different capacities and coping strategies in order to design effective long-term adaptation, mitigation and resilience programmes.

2.4 Climate change impact on women's livelihoods

> Gender inequalities intersect with climate risks and vulnerabilities. Women's historic disadvantages – their limited access to resources, restricted rights, and a muted voice in shaping decisions – make them highly vulnerable to climate change.
>
> (Human Development Report 2007/08 [UNDP, 2007])

Gender discrimination is one of most striking dimensions and indicators of inequality; however, it is often overlooked in climate change-related discussions and interventions. Although climate change affects everyone, it is not gender-neutral. Climate change magnifies existing inequalities, reinforcing the disparity between women and men in their vulnerability towards and capacity to cope with climate change (UNDP, 2007). Women, as the majority of the world's poor, are the most vulnerable to the effects of climate change (WEDO, 2007). Three reasons are provided for the increased severity of climate change effects on women than men. These are: biological and physical differences between men and women; pre-existing social norms; and values that have intensified a new form of gender discrimination (UN WomenWatch, 2009).

The gender gap provides reasonable grounds for the expectation that women and men will generally be affected differently by the effects of climate change and will therefore respond to and benefit differently from climate protection and adaptation measures (World Bank, 2011). Women are more vulnerable both to short-term recurring climatic events and long-term climate-induced changes because of gender differences in socially constructed roles and responsibilities that affect mobility, social networks and access to information and local institutions, as well as access to, control and ownership of assets (BCAS, 2010).

In developing countries like Bangladesh, more women die during natural disasters compared to men, because they are not adequately warned, cannot swim well or cannot leave the house alone due to cultural constraints (UNFCCC, 2005). Moreover, lower levels of education reduce the ability of women and girls to access information, including early warning systems and resources, or to make their voices heard. Bangladesh often experiences natural hazards and, over time, most water-related hazards will be exacerbated due to climate-induced effects (Ahmed, 2005). An Oxfam International's report (2009) on the impacts of the 2004 Asian tsunami on women raised alarm about gender imbalances, since the majority of those killed and among those least able to recover were women. In Ache, Indonesia, for example, more than 75 per cent of those who died were women, resulting in a male-female ratio of 3:1 among the survivors. Cyclonic storm surge is a common natural hazard in coastal Bangladesh. Following the cyclone and flood disasters of 1991, it was revealed that among women aged 20–44, the death rate was 71 per 1000, compared to 15 per 1000 for men (UNEP, 2005). During Cyclone Sidr, which hit in 2007, the gender gap in mortality rates was 5:1, specifically implying that there may be cultural reasons as to why women were reluctant to use cyclone shelters. During cyclones and floods, women, including adolescent girls, suffer the most as they find it difficult to quickly move to safety before and after any disaster hits (WEDO, 2008).

In developing countries, women are heavily engaged in resource-dependent activities such as agriculture and forest management (Dankelman & Davidson, 1988). Climate variability directly affects natural resource bases and, thus, resource-dependent activities. Therefore, women are likely to be affected more severely than men due to their overwhelming dependency and low technical knowledge regarding how to adapt to unexpected situations (Baten & Khan, 2010). Climate change, which reduces crop yields and food production particularly in developing countries, affects women's livelihood strategies and food security, and therefore, their right to food. Rural women's lack of access to and control over natural resources, technologies and credit mean that they have fewer resources to cope with seasonal and episodic weather and natural disasters. Consequently, traditional roles are reinforced, girls' education suffers, and women's ability to diversify their livelihoods is diminished (Masika, 2002).

Every year, more than 100 million women and girls are affected by natural disasters (UNISDR, 2012). Many will bear the brunt of recurring cyclones, sea level rise, floods, storms or droughts. On 13 October 2012, UNISDR declared that women and girls were the focus of that year's theme for the International Day for Disaster Reduction. The UNDP, DFID, UNEP, World Bank and other organisations have also taken some initiatives to address women's vulnerability in some regions in Asia-Pacific and Africa. In Bangladesh, though, very limited initiatives have been undertaken. As such, it is very difficult to find specific studies about women's vulnerabilities, adaptation and coping mechanisms in coastal areas of Bangladesh. However, some studies have been conducted in other areas and other dimensions of research that give some ideas about gender inequalities in relation to climate change impacts in Bangladesh.

Under the prevailing social and economic circumstances, Bangladeshi women are lagging far behind their male counterparts. Women's and men's responses to these crisis situations, as well as their abilities to cope with them, to a very large extent reflect their status, roles and positions in society: because of gender-based inequalities, girls and women are typically at higher risk than men and boys (Chew & Ramdas, 2005). In Bangladesh, women are adapting to climate change in traditional ways. The majority of them, however, do not have enough understanding about the effectiveness of each of the preparedness measures available, as well as their limitations. Often these measures do not help them because of the magnitude of disasters (WEDO, 2008).

As reported by the gender disaster workshop in Ankara (DAW-UNISDR, 2001),

Women's human rights are not comprehensively enjoyed throughout the disaster process. Economic and social rights are violated in disaster processes if

mitigation, relief, and reconstruction policies do not benefit women and men equally. The right to adequate health care is violated when relief efforts do not meet the needs of specific physical and mental health needs throughout their life cycle, in particular when trauma has occurred. The right to security of persons is violated when women and girls are victims of sexual and other forms of violence while in relief camps or temporary housing. Civil and political rights are denied if women cannot act autonomously and participate fully at all decision-making levels in matters regarding mitigation and recovery.
(DAW-UNISDR, 2001, Summary of Debate, Section B, p. 10)

While awareness of climate change and its impacts has risen dramatically since the original agreement of COP 15 in 2009, some key areas have been missing from the debate, and the gendered nature of climate change is one of them. International climate change policy makers have long neglected the gender dimension of climate change, as well as other social and political factors. Even though vulnerability to environmental degradation and natural hazards have been linked to poverty, gender and other social dimensions, there has been little análysis on both the positive and negative impacts of climate change on gender and social relations. Overall, women are poorly represented in planning and decision-making processes in climate change policies, limiting their capacity to engage in political decisions that can impact upon their specific needs and vulnerabilities (Climate Alliance, 2005).

Given the gender differential in vulnerability, it is important to have social assessments and institutional analyses that include gender-based experiences in collective actions and support from local institutions and social networks for developing inclusive strategies for increased climate resilience. Gender-sensitive analysis is also important in planning for full and equitable recovery in the case of frequent climatic events, such as floods and cyclones, whose frequency and intensity are expected to rise with climate change. Furthermore, gender-sensitive analysis is important in ensuring women's participation in long-term climate change adaptation strategies, which might have been previously constrained due to the traditional social norms and values in Bangladesh. Keeping the above discussions in mind, this study addresses climate change as a gendered issue by examining climate change vulnerabilities from the perspective of rural women, the types and level of their vulnerability, and makes recommendations as to how gender disparities may be addressed in the context of a climate changing future.

Notes

1 Purchasing power parity (PPP) is a component of some economic theories and is a technique used to determine the relative value of different currencies.
2 Grameen Bank has offered credit almost exclusively to poor women from households that own no farmland or other significant assets. The programme is 'minimalist,' specialising in the delivery of small loans for a short duration at a reasonable rate of interest. All borrowers must make a commitment to a compulsory saving regime, which acts as a form of loan default insurance programme. See http://www.inquiriesjournal.com/articles/1451/rethinking-microcredit-in-bangladesh-does-grameen-bank-serve-the-neoliberal-agenda; http://journals.sagepub.com/doi/abs/10.1177/0921374007088053; https://newint.org/books/reference/world-development/case-studies/poverty-microcredit-grameen-bank/

References

ACCCRN. (2012). *Asian cities climate change resilience network brochure*. Asian Cities Climate Change Resilience Network (ACCCRN), supported by the Rockefeller Foundation, Bangkok.

ADB (2001). Women in Bangladesh: Country Briefing Paper. Programs Department (West), Asian Development Bank (ADB), Manila.

Ahmed, A. U. (2005). Adaptations options for managing water related extreme events under climate change regime: Bangladesh perspective. In M. M. Q. Mirza, & Q. K. Ahmed (Eds.), *Climate change and water resources in South Asia* (pp. 255–278). Leiden: Balkema Press.

Alam, M., Nishat, A., & Siddiqui, S. M. (1998). Water resources vulnerability to climate change with special reference to inundation. In S. Huq, Z. Karim, M. Asaduzzaman, & F. Mahtab (Eds.), *Vulnerability and adaptation to climate change for Bangladesh* (pp. 21–38). Dordrecht: Kluwer Academic Publishers.

Alam, K., Rahman, A., Farok, O., Kabir, F., & Fatema, N. (2007). *Drowning sand and the Holy banana tree: The tale of people with disability and their neighbors coping with sharbanasha floods in the Brahmaputra-Jamuna chars of Bangladesh* (57 pp.). Dhaka: Handicap International.

Asaduzzaman, M., Rahman, M. H., & Jahan, H. (2004). Adoption of selected homestead agricultural technologies by rural women in Madhupur Upazila under Tangail District. *Progressive Agriculture, 15*(1), 57–63.

Asaduzzaman, M., Reazuddin, M., & Ahmed, A. U. (Eds.) (1997). *Global climate change: Bangladesh episode*. Dhaka: Department of Environment, Government of the People's Republic of Bangladesh.

Baida, S. K., & Shrestha, R. K. (2007). *Climate profile and observed climate change and climate variability in Nepal (Final draft)*. Katmandu: Department of Hydrology and Meteorology, Ministry of Science, Technology and Environment, Government of Nepal.

Baten, M. A. & Khan, N. A. (2010). Gender Issue in Climate Change Discourse: Theory versus Reality. Unnayan Onneshan-The Innovators: Centre for Research and Action on Development, Dhaka.

Bayes, A. (2015). *Beneath the Surface: Microcredit and Women's Empowerment*. Kobe: The Virtual Library of Microcredit and Microfinance, the Global Development Research Center (GDRC).

BBS. (2013a). *Literacy assessment survey (LAS)-2011*. Dhaka: Bangladesh Bureau of Statistics, Statistics and Informatics Division (SID), Ministry of Planning, Government of the People's Republic of Bangladesh.

BBS. (2013b). *Literacy assessment survey (LAS)-2011*. Dhaka: Bangladesh Bureau of Statistics, Statistics and Informatics Division (SID), Ministry of Planning, Government of the People's Republic of Bangladesh.

BBS. (2013c). *Gender statistics of Bangladesh 2012*. Dhaka: Bangladesh Bureau of Statistics, Statistics and Informatics Division (SID), Ministry of Planning, Government of the People's Republic of Bangladesh.

BBS. (2015). *Highlights*. Dhaka: Bangladesh Bureau of Statistics, Statistics and Informatics Division (SID), Ministry of Planning, Government of the People's Republic of Bangladesh. Retrieved from www.bbs.gov.bd/home.aspx

BCAS. (2010). *Gender and climate change issues in the South Central and South West Coastal regions of Bangladesh*. Dhaka: Bangladesh Centre for Advanced Studies (BCAS).

BCAS-RA-Approtech. (1994). Vulnerability of Bangladesh to climate change and sea level Rise: concepts and tools for calculating risk in integrated coastal zone management. In

Four volumes (Summary report, main reports and institutional report). Dhaka: Bangladesh Centre for Advanced Studies (BCAS), Resource Analysis (RA), and Approtech Consultants Ltd.

BGMEA. (2015). Bangladesh Garment Manufacturers and Exporters Association. Retrieved from www.bgmea.com.bd/home/about/Strengths

Chew, L., & Ramdas, K. (2005). *Caught in the storm: The impact of natural disasters on women*. San Francisco: The Global Fund for Women.

Choudhury, A. M., Neelormi, S., Quadir, D. A., Mallick, S., & Ahmed, A. U. (2005). Socio-economic and physical perspectives of water related vulnerability to climate change: Results of field study in Bangladesh. *Science and Culture (Special Issue)*, *71*(7–8), 225–238.

Climate Alliance. (2005). *Climate alliance 2004/05 annual report* (74 pp.). Frankfurt am Main: Climate Alliance.

CPD and MJF. (2014). *Estimating women's contribution to the economy: The case of Bangladesh*. Presented at the dialogue on 'how much women contribute to the Bangladesh economy', results from an empirical study organised by Centre for Policy Dialogue (CPD) in partnership with Manusher Jonno Foundation (MJF) on May 05, 2015, Dhaka.

Dankelman, I., & Davidson, J. (1988). *Women and environment in the third world: Alliances for the future*. London: Eartscane Publications Ltd (in association with IUCN).

DAW-UNISDR. (2001, November 6–9). *Environmental management and the mitigation of natural disasters: A gender perspective*. Report of the expert group meeting, Ankara, Turkey.

Deb, U. K., Khaled, N., Al Amin, M., & Nabi, A. (2009, January 2). *Climate change and rice production in Bangladesh: Implications for R&D strategy*. Paper presented at the special conference on "climate change and Bangladesh development strategy: Domestic tasks and international cooperation" organised by Bangladesh Paribesh Andolan (BAPA) and Bangladesh Environment Network (BEN), Dhaka.

EM-DAT-The International Disaster Database. (2015). *Center for research on the epidemiology of disaster*. Retrieved from www.emdat.be/database

Ericson, J. P., Vorosmarty, C. J. Dingman, S. L., Ward, L. G., & Meybeck, M. (2006). Effective Sea-level Rise and Deltas: Causes of Change and Human Dimension Implications. *Global Planet Change*, *50*(1–2), 63–82.

FAO. (2007). *Climate variability and change: Adaptation to drought in Bangladesh: A resource book and training guide*. Prepared by Asian Disaster Preparedness Center and Food and Agriculture Organization of the United Nations.

Ghatak, M., Kamal, A., & Mishra, O. P. (2012, October 9–10). Background paper: Flood risk management in South Asia. In *Proceedings of the SAARC workshop on flood risk management in South Asia*, Islamabad, Pakistan.

Gibson, S., Mahmud, S., Toufique, K. A., & Turton, C. (2004). *Breaking new ground: Livelihood choices, opportunities and tradeoffs for women and girls in rural Bangladesh*. Somerset: The IDL Group.

Global Assessment Report (GAR). (2011). *Global assessment report on disaster risk reduction 2011: Revealing risk, redefining development*. New York: UN Publications.

GoB (2008a). *Cyclone Sidr in Bangladesh: Damage, Loss and Needs Assessment for Disaster Recovery and Reconstruction*. A Report Prepared by the Government of Bangladesh, Assisted by the International Development Community with Financial Support from the European Commission, Dhaka.

GoB (2008b). *Millennium Development Goals: Bangladesh Progress Report 2008*. Bangladesh Planning Commission, Government of the People's Republic of Bangladesh, Dhaka.

Handmer, J., Honda, Y., Kundzewicz, Z. W., Arnell, N., Benito, G., Hatfield, J., Mohamed, I. F., Peduzzi, P., Wu, S., Sherstyukov, B., Takahashi, K., & Yan, Z. (2012). Changes in impacts of climate extremes: Human systems and ecosystems. In C. B. Field, V. Barros, T. F. Stocker, D. Qin, D. J. Dokken, K. L. Ebi, M. D. Mastrandrea, K. J. Mach, G.-K. Plattner, S. K. Allen, M. Tignor, & P. M. Midgley (Eds.), *Managing the risks of extreme events and disasters to advance climate change adaptation.* A special report of Working Groups I and II of the Intergovernmental Panel on Climate Change (IPCC) (pp. 231–290). Cambridge and New York: Cambridge University Press.

Haque, U., Hashizume, M., Kolivras, K. N., Overgaard, H. J., Das, B., & Yamamoto, T. (2012). Reduced death rates from cyclones in Bangladesh: What more needs to be done? *Bulletin of the World Health Organization (WHO), 90*(2), 150–156. Retrieved from http://thinkprogress.org/climate/2012/10/14/1009121/science-of-global-warming-impacts-guide/. www.un.org/womenwatch/feature/climate_change/

IFAD. (2015). *Investing the rural people in Bangladesh.* Rome: International Fund for Agricultural Development (IFAD).

IPCC. (2007a). *Climate change 2007: The physical science basis.* Contribution of Working Group I to the fourth assessment report of the Intergovernmental Panel on Climate Change IPCC (IPCC) (996 pp.). Cambridge and New York: Cambridge University Press.

IPCC (2007b). *Climate change 2007: Impacts, adaptation and vulnerability. contribution of Working Group II to the Fourth Assessment Report of the Intergovernmental Panel on Climate Change (IPCC)* (976 pp.). Cambridge: Cambridge University Press, Cambridge.

Islam, A. K. M. N. (2008). *SIDR cyclone in Bangladesh: End of the course project.* Submitted to the National Institute of Disaster Management (NIDM), New Delhi, for 4th Online Course on Comprehensive Disaster Management, New Delhi.

Islam, N., & Sultana, N. (2006). The status of women in Bangladesh: Is the situation really encouraging? *Research Journal of Social Sciences, 1*(1), 56–65.

Islam, R. (2009). *Climate change induced disasters and gender dimensions: Perspective Bangladesh.* Costa Rica: Peace and Conflict Monitor, University of Peace.

Karim, Z., Hussain, S. G., & Ahmed, A. U. (1998). Climate change vulnerability of crop agriculture. In S. Huq, Z. Karim, M. Asaduzzaman, & F. Mahtab (Eds.), *Vulnerability and adaptation to climate change for Bangladesh* (pp. 39–54). Dordrecht: Kluwer Academic Publishers.

Khan, M.S.A. (2008). Disaster preparedness for sustainable development in Bangladesh. *Disaster Prevention and Management, 17*(5), 662–671.

Khan, A. E., & Islam, M. I. (2011, June 5). *Water salinity and maternal health.* Article published in *The Daily Star,* Dhaka.

Khatun, F., & Islam, A. K. M. N. (2010). *Policy agenda for addressing climate change in Bangladesh: Copenhagen and beyond* (Occasional Paper No. 88). Dhaka: Centre for Policy Dialogue (CPD).

Masika, R. (Ed.) (2002). *Gender, development, and climate change.* Oxfam focus on gender series (115 pp.). Oxford: Oxfam GB.

Mirza, M. M. Q., & Paul, S. (1992). *Natural disaster and environment in Bangladesh* (in Bengali). Dhaka: Centre for Environmental Studies and Research.

MoEF. (2008). *Bangladesh climate change strategy and action plan 2008.* Dhaka: Ministry of Environment and Forests (MoEF), People's Republic of the Government of Bangladesh (GoB).

Neumayer, E. & Plümper, T. (2007). The gendered nature of natural disasters: The impact of catastrophic events on the gender gap in life expectancy, 1981–2002. *Annals of the Association of American Geographers, 97*(3), 551–566.

NIPORT. (2015). *Bangladesh demographic and health survey 2014: Key indicators.* Dhaka: National Institute of Population Research and Training (NIPORT), Ministry of Health and Family Welfare, People's Republic of the Government of Bangladesh.

Oxfam International. (2009). In the wake of the Tsunami: An evaluation of Oxfam International's response to the 2004 Indian Ocean Tsunami. In *Oxfam International Tsunami fund final evaluation series: Summary report.* Oxford: Oxfam International.

Peduzzi, P., Chatenou, B., Dao, H., De Bono, A., Herold, C., Kossin, J., Mouton, F., and Nordbeck, O. (2012). Global trends in tropical cyclone risk. *Nature Climate Change, 2*(4), 289–294.

PRDI. (2015a). *Increasing salinity threatens: Productivity of Bangladesh.* Dhaka: Participatory Research and Development Initiative (PRDI).

PRDI. (2015b). *Salinity threatens the world heritage: The Sundarban of Bangladesh.* Dhaka: Participatory Research and Development Initiative (PRDI). Retrieved from http://preventionweb.net/go/8203

Rashid, H. E. (1991). *Geography of Bangladesh* (second revised edition). Dhaka: The University Press Ltd.

Rashid, M. M., Hoque, A. K. F., & Iftekhar, M. S. (2004). Salt tolerances of some multipurpose tree species as determined by seed germination. *Journal of Biological Sciences, 4*(3), 288–292. Retrieved from http://preventionweb.net/go/8199

Romm, J. (2012). *An illustrated guide to the science of global warming impacts: How we know inaction is the gravest threat humanity faces.* Published in the climate progress website on October 14, 2012. Retrieved from http://thinkprogress.org/climate/2012/10/14/1009121/science-of-global-warming-impacts-guide/.

Sarwar, M. G. M. (2005). *Impacts of sea level rise on the coastal zone of Bangladesh* (Unpublished Master thesis). Submitted to the Lund University, Sweden.

Sen, G. (2000). Engendering poverty alleviation: Challenges and opportunities. In S. Razavi (Ed.), *Gendered poverty and well-being* (291 pp.). Oxford: Blackwells.

Stern, N. (2006). *Report of the stern review: The economics of climate change.* London: HM Treasury, Government of UK.

The Daily Star. (2015, May 06). Women's contribution to development unrecognised. News published in *The Daily Star*, Dhaka.

UN WomenWatch. (2009). *Fact sheet: Women, gender equality and climate change.* United Nations Inter-Agency Network on Women and Gender Equity (IANWGE). Retrieved from http://www.un.org/womenwatch/feature/climate_change/.

UNDP. (2002, June 17–19). *Integrating disaster reduction with adaptation to climate change.* Synthesis of UNDP expert group meeting, Havana, Cuba.

UNDP. (2007). *Human development report 2007–08: Fighting climate change: Human solidarity in a divided world.* New York: United Nations Development Programme, Palgrave Macmillan.

UNDP. (2013). *Human development report 2013, the rise of the South: Human progress in a diverse world.* Explanatory note on 2013 HDR composite indices. New York: United Nations Development Programme (UNDP).

UNDP. (2014). *Human development report 2014, sustaining human progress: Reducing vulnerabilities and building resilience.* New York: United Nations Development Programme (UNDP).

UNEP. (1989). *United Nations Environment Programme.* Retrieved from www.grida.no

UNEP. (2005). *Mainstreaming gender in environmental assessment and early warning.* Nairobi: United Nations Environment Programme (UNEP).

UNFCCC. (2005, December). *Global warming: Women matter*. United Nations Framework Convention on Climate Change (UNFCCC) COP Women's Statement.

UNICEF. (2001). *Rural adolescent girls in Bangladesh*. Dhaka: United Nations Children's Fund (UNICEF).

UNICEF. (2010). *Women and girls in Bangladesh: Key statistics*. Dhaka: United Nations Children's Fund (UNICEF).

UNICEF. (2015). *Cyclone Aila*. Retrieved from UNICEF Bangladesh webpage: www.unicef.org/bangladesh/4926_6202.htm

UNICEF and BBS. (2010). Bangladesh: Monitoring the situation of children and women-multiple indicator cluster survey 2009. In *Progotir Pathey 2009 volume I: Technical report*. Dhaka: United Nations Children's Fund (UNICEF), UN and Bangladesh Bureau of Statistics (BBS), Ministry of Planning, Government of Bangladesh.

UNIFEM. (2008). *Concept paper on climate change and gender*. New York: United Nations Development Fund for Women (UNIFEM).

UNISDR. (2012). *Global assessment report on disaster risk reduction: Revealing risk, redefining development*. Geneva: The United Nations Office for Disaster Risk Reduction (UNISDR).

Unnikrishnan, A. S., Kumar, R. K., Fernandes, S. E., Michael, G. S., & Patwardhan, S. K. (2006). Sea level changes along the Indian coast: Observations and projections. *Current Science India, 90*(3), 362–368.

Warrick, R. A., & Ahmad, Q. K. (Eds.) (1996). *The implications of climate and sea-level change for Bangladesh* (415 pp.). Dordrecht: Kluwer Academic Publishers.

WEDO (2007). Changing the Climate: Why Women's Perspective Matter. Fact Sheet. Women's Environment and Development Organization (WEDO). New York.

WEDO. (2008). *Gender, climate change and human security: Lessons from Bangladesh, Ghana and Senegal*. Report prepared by the Women's Environment and Development Organization (WEDO) with ABANTU for development in Ghana, action aid Bangladesh and ENDA in Senegal.

WHO (2011). World Health Statistics 2011. Geneva: World Health Organization.

World Bank. (2000). *Bangladesh: Climate change and sustainable development* (Report No. 21104-BD, 95 pp.). Dhaka: Rural Development Unit, South Asia Region, The World Bank.

World Bank. (2011). *Gender and climate change: Three things you should know* (p. 7). Washington, DC: The Social Development Department, The World Bank.

World Bank. (2013, June 19). *Warming climate to hit Bangladesh hard with sea level rise, more floods and cyclones*. Washington, DC: World Bank.

World Bank. (2015). *The World Bank DataBank*. Retrieved from http://data.worldbank.org/indicator/EN.POP.DNST

3 Vulnerability and the Sustainable Livelihood Framework

Methodological considerations

This chapter provides an outline of the theoretical framework and research methods used in this study. Under the theoretical framework, it firstly provides an understanding of the Sustainable Livelihood Framework (SLF) that is used in this book to assess the livelihood of women in the context of climate variability and extremes. Secondly, this chapter briefly presents the concept of 'vulnerability' and how this was assessed in respect to different climate change effects. The chapter then describes the research approach, the selection of the study area, the methods of data collection and the data analysis. Under data analysis, the chapter specifically explains the two methods of vulnerability measurements, the Livelihood Vulnerability Index (LVI) and the IPCC-LVI, that are used to estimate the vulnerability index for women in the study area.

3.1 The Sustainable Livelihood Framework

The word 'livelihood' can be defined in many ways. These definitions attempt to capture the broad notion of livelihoods, which are understood as "the capabilities, assets (including both material and social resources) and activities required for a means of living" (Chambers & Conway, 1992, p. 6). The concept of livelihood extends to

> the command an individual, family, or other social group has over an income and/or bundles of resources that can be used or exchanged to satisfy its needs. This may involve information, cultural knowledge, social networks and legal rights, as well as tools, land or other physical resources.
>
> (Blaikie, Cannon, Davis, & Wisner, 1994, p. 23)

The goal of any livelihoods strategy is to develop self-reliance. According to Ellis (2000), the 'livelihoods' concept is a realistic recognition of the multiple activities, in which households engage to ensure their survival and improve their wellbeing.

Acknowledging the pressure caused by enormous population, Chambers and Conway (1992) developed the idea of 'sustainable livelihoods' with the intention of enhancing the efficiency of development cooperation programmes of international agencies worldwide. According to them, "a livelihood is sustainable when it can cope with and recover from stresses and shocks and maintain or enhance its capabilities and assets both now and in the future, while not undermining the natural resource base" (Chambers & Conway, 1992, p. 9). The analytical framework of the sustainable livelihood was first revealed by Ian Scoones as that which investigates the role of livelihood resources and institutional processes within the sustainable livelihood approach (Scoones, 1998). Later, this understanding of the Sustainable Livelihood Framework was modified and elaborated upon by various development agencies. In 1999, the UK government's Department for International Development (DFID) introduced a new version of the Sustainable Livelihood Framework, which is often called the DFID framework in livelihood literature (DFID, 1999). This new framework added important components, such as a 'vulnerability context,' which introduces more elaborate discussion on this framework. The framework emphasises the main factors that influence livelihoods, and it allows us to know the inner relationships among these factors. The DFID's framework is centred on people and does not aim to show the model of reality; rather, it helps the stakeholders to engage in 'structured and coherent debate' about the many factors that influence livelihoods, their relative value and the way in which they interact (DFID, 1999, p. 1). The main feature of this framework is obviously 'livelihood assets,' which is comprised of five important capitals that are discussed later.

The DFID sustainable livelihood framework has been found to be a useful tool for livelihood analysis, which is represented as a methodological framework (Ellis, 2000). This framework was also popularised by other international NGOs, such as Cooperative for Assistance and Relief Everywhere (CARE), the International Union for Conservation of Nature (IUCN), OXFAM, the UNDP, etc. Consequently, the present study uses the concept of the SLF to examine the livelihood of women in the coastal areas of Bangladesh, and is discussed further under the following sub-headings.

3.1.1 Core concepts of the SLF

The SLF has six core principles that should not be compromised, though the Framework is flexible in its application. DFID describes these core concepts as:

- **People-centred**: People rather than the resources are used as the priority concern in this livelihoods approach, since problems associated with development are often rooted in adverse institutional structures that are impossible to overcome through simple asset creation.
- **Holistic**: A holistic view is aspired to which understands the stakeholder's livelihoods as a whole, with all facets included. It is non-sectoral and applicable across geographical areas and social groups. It recognises multiple influences on and multiple actors involved with people. Essentially, it seeks to achieve multiple livelihood outcomes that are determined by people.

- **Dynamic**: Just as people's livelihoods and the institutions that shape them are highly dynamic, so is the approach to learning from changes. It helps people in mitigating negative impacts, whilst supporting positive effects.
- **Building on strengths**: A central issue of the approach is the recognition of everyone's inherent potential for his/her removal of constraints and realisation of potentials. This will contribute to the stakeholders' robustness and ability to achieve their own objectives.
- **Macro-micro links**: Development activity tends to focus at either the macro or the micro level, whereas the SLF tries to bridge this gap in stressing the links between the two levels.
- **Sustainability**: A livelihood can be classified as 'sustainable' when it is resilient in the face of external shocks and stresses, when it is not dependent upon external support, when it is able to maintain the long-term productivity of natural resources and when it does not undermine the livelihood options of others (Sneddon, 2000; Kollmair & Gamper, 2002).

3.1.2 Modelling the SLF

Figure 3.1 illustrates the SLF that serves as an instrument for the investigation of people's livelihoods. The framework depicts stakeholders as operating in a

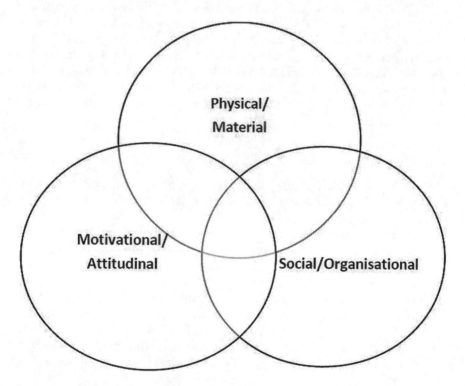

Figure 3.1 The interlocking nature of vulnerability

context of vulnerability, within which they have access to certain assets. These gain their meaning and value through the prevailing social, institutional and organisational environment (transforming structures and processes). This context decisively influences the livelihood strategies that are open to people in pursuit of their self-defined beneficial livelihood outcomes. The framework consists of a number of key elements, which are:

i)   Livelihood assets
ii)   Vulnerability context
iii) Transforming structures and processes (policies, institutions, and processes)
iv)   Livelihood strategies
v)   Livelihood outcomes

i) Livelihood capital/assets

The livelihoods approach is concerned first and foremost with people. So an accurate and realistic understanding of people's strengths (here called "assets" or "capital") is crucial in order to analyse how they endeavour to convert their assets into positive livelihood outcomes (Bebbington, 1999). Assets refer to the resource base of people. Assets are often represented as a pentagon in the SLF, consisting of the following five categories: labour with different skills (human capital); natural resources (natural capital); monetary resources (financial capital); social networks of various kinds (social capital); and physical reproducible goods (physical capital). Since the importance of the single categories varies in association with the local context, the asset pentagon offers a valuable tool for visualising these settings and demonstrating dynamic changes over time through constantly shifting shapes of the pentagon (Kollmair & Gamper, 2002).

The asset pentagon lies at the core of the livelihoods framework 'within' the vulnerability context. The pentagon was developed to enable information about people's assets to be presented visually, thereby bringing to life important inter-relationships between the various assets. The shape of the pentagon can be used to show schematically the variation in people's access to assets (DFID, 1999). The centre point of the pentagon, where the lines meet, represents zero access to assets while the outer perimeter represents maximum access to assets. In that way a complete pentagon represents the availability of adequate resources for livelihood operation, while a squeezed pentagon represents the opposite. On the basis of resource availability, different shapes of pentagons can be drawn for different communities or different social groups within the same communities. These five capitals form an 'asset pentagon' which lies at the core of the SLF. These categories are discussed here:

i)   **Human capital**: Human capital represents the skills, knowledge, ability to labour and good health that together enable people to pursue different livelihood strategies and achieve their livelihood objectives. At a household level, human capital is a factor of the amount and quality of labour available; this varies according to household (HH) size, skill levels, leadership potential, health status, etc.

ii) **Natural capital**: Natural capital is the term used for the natural resource stocks from which resource flows and services (e.g. food cycling, disaster protection) useful for livelihoods are derived. There is a wide variation in the resources that make up natural capital, from intangible public goods, such as the atmosphere and biodiversity, to divisible assets used directly for production (trees, land, etc.). Within the SLF, the relationship between natural capital and the *vulnerability context* is particularly close. Many of the shocks that devastate the livelihoods of the poor are themselves natural processes that destroy natural capital (e.g. cyclones that destroy forests, floods and saline water that destroy agricultural land) and seasonality is largely due to changes in the value or productivity of natural capital over the year.

iii) **Social capital**: In the context of the SLF, social capital is taken to mean the social resources upon which people draw in pursuit of their livelihood objectives. These are developed through networks and connectedness, membership in more formalised groups and social relations and trust, reciprocity and access to wider institutions in society. People use these networks to reduce risks, access services, protect themselves from deprivation and to acquire information on lower transection costs.

iv) **Physical capital**: Physical capital comprises the 'basic infrastructure' and 'producer goods' needed to support livelihoods. The components of 'basic infrastructure' and 'producer goods' include affordable transport; secure shelter and buildings; adequate water supply and sanitation; clean, affordable energy; and access to information.

v) **Financial capital**: Financial capital denotes the financial resources that people use to achieve their livelihood objectives. The two main sources of financial capital are 'available stocks' and 'regular inflows of money.' Available stocks can be held in several forms, such as cash, bank deposits or liquid assets, such as livestock and jewellery, or resources obtained from credit-providing institutions. Regular inflows of money include earned income, pensions, other transfers from the state and remittances.

McLeod (2001) proposes to include analysis of political capital in livelihood assets. This goes beyond social capital in that an individual's stock of political capital will determine their ability to influence policy and the processes of government. An understanding of political capital is important in determining the ability of households and individuals to claim rights. However, it has limited implications for vulnerable groups, like women in Bangladesh, because of low organisational participation and low decision-making capacity in any sphere of their lives.

Assets on which households or individuals draw to build their livelihoods are at the centre of the SLF. Therefore, the SLF requires a realistic understanding of these assets in order to identify what opportunities they may offer or where constraints may lie (Rakodi, 2002). The present study has also considered the assets of the SLF as the main focus of the livelihood assessment.

ii) Vulnerability context

The vulnerability context forms the external environment in which people exist and how their asset status can be directly impacted (Devereux, 2001). It is comprised of trends (i.e. demographic trends, resource trends, trends in governance), shocks (i.e. human, livestock or crop health shocks; natural hazards, like cyclone or salinity increase; economic shocks; situations of political unrest; conflicts in the form of national or international wars) and seasonality (i.e. seasonality of prices, products or employment opportunities) and represents the part of the framework that lies furthest outside of a stakeholder's control. Not all trends and seasonality must be considered as negative; they can move in favourable directions too. The 'vulnerability' concept is discussed further in section 3.2 of this chapter.

iii) Transforming structures and processes

Transforming structures and processes occupy a central position in the SLF. Transforming structures and processes represent the institutions, organisations, policies and legislation that shape livelihoods. They are of central importance as they operate at all levels and effectively determine access, terms of exchange between different types of capital and returns to any given livelihood strategy (Kollmair & Gamper, 2002). 'Structure' and 'processes' can be described in similar terms as the 'hardware' and 'software' of a computer. Structures can be described as the 'hardware' (private and public organisations) that implement policy and legislation, deliver services, purchase, trade and perform all manner of other functions that affect livelihood (DFID, 2000). Complementary to these structures, processes constitute the 'software' determining the way in which structures and individuals operate and interact. There are many types of overlapping and conflicting processes that operate at a variety of levels and, like software, they are crucial and complex (DFID, 1999).

iv) Livelihood strategies

Livelihood strategies comprise the range and combination of activities and choices that people undertake in order to achieve their livelihood goals. They have to be understood as a dynamic process in which people combine activities to meet their various needs at different times and on different geographical or economical levels. People's direct dependence on asset status and transforming structures and processes becomes clear through the position they occupy within the framework. When considering livelihood strategies and issues connected to the SLF in general, it is important to recognise that people compete (for jobs, markets, natural resources, etc.), which makes it difficult for everyone to achieve simultaneous improvements in their livelihoods. The poor are themselves a very heterogeneous group, placing different priorities in a finite, and therefore highly disputed, environment. An application of the SLF offers the advantage of being sensitive to such issues in a differentiated manner (DFID, 1999; Kollmair & Gamper, 2002).

v) Livelihood outcomes

Livelihood outcomes are the achievements of livelihood strategies, such as more income (e.g. cash), increased wellbeing (e.g. non material goods like self-esteem, health status, access to services, sense of inclusion), reduced vulnerability (e.g. better resilience through an increase in asset status), improved food security (e.g. increase in financial capital in order to buy food) and a more sustainable use of natural resources (e.g. appropriate property rights). Tangible outcomes allow for an understanding of the 'output' or application of current factors within the livelihood framework; they demonstrate what motivates stakeholders to act as they do and what their priorities are. They might give us an idea of how people are likely to respond to new opportunities and which performance indicators should be used to assess support activity. Livelihood outcomes directly influence the assets and dynamically change their form of pentagon in terms of the weighting given to specific categories, offering a new starting point for other strategies and outcomes (DFID, 1999; Kollmair & Gamper, 2002).

3.1.3 Application of the Sustainable Livelihood Framework around the world and its relevance to this study

The use of the SLF in the late 1990s grew exponentially. Many organisations used its principles and developed their own frameworks. The UK Department for International Development (DFID) made an enormous investment in this area from 1997 to 2002. This investment led to the significant implementation of the framework within the wider international community (Overseas Development Institute, 2003). The SLF is now being applied and adapted to different development challenges, particularly community-driven development, making markets work for the poor, food security, climate change and disaster risk reduction (Institute of Development Studies, 2011). It is now the norm to use the term 'livelihood' instead of 'employment.' The language has changed significantly in the last 15 years – from thinking about an 'adequate' and 'decent' standard of living to 'livelihoods strengthening,' 'livelihoods diversification' and beyond that, linking to discourses on climate change, resilience and power – which present important steps in moving forward (IDS, 2011).

From 1997, the DFID integrated the framework in its programme for development cooperation. The framework is used in many ways all around the world. This is seen in: the Western Orissa Rural Livelihoods Project (WORLP) in India; the livelihoods-centred approaches to disaster risk reduction in Nepal; adaptation strategies among pastoral and agro-pastoral communities in Ethiopia and Mali; and, the Khanya-aicdd project in South Africa for community-driven development, including key governance issues (IDS, 2011). However, sustainable livelihood frameworks are also being used with some modification by agencies like CARE International, Practical Action, the Canadian International Development Agency (CIDA) and some other NGOs.

Modified sustainable livelihood frameworks are also being used to improve the livelihoods of people living in poverty and in the reduction of climate change vulnerabilities. Some instances of these are: the Adaptation to Climate Change in Tajikistan (ACCT) project; the Climate Vulnerability and Capacity Analysis (CVCA) process project in Ghana; and the Practical Action project in Bangladesh (IDS, 2011). It has been proven that the sustainable livelihood framework can be used successfully worldwide to address the livelihoods of targeted people (e.g. poor, women, people of disaster-prone areas). The capacity of this framework to assess the level of vulnerability, adaptability and resilience makes it a useful tool for livelihood studies, which can help with advancing the policies aimed at sustainable development. Consequently, based on the research theme and the objectives that were outlined in Chapter 1, the present study finds it a useful and appropriate model to be applied to the assessment of the livelihood of women in a climate vulnerable context in Bangladesh. Therefore, the SLF has been chosen as the main research model to be utilised in the present study.

3.2 Understanding and assessing vulnerability

The concept of vulnerability was rooted in the study of natural hazards (Hewitt, 1983). More recently, it has become central concept in a variety of research contexts, including natural hazards and disaster management, ecology, public health, poverty and development, rural livelihoods and famine, sustainability science, land use change and climate impacts and adaptation (Füssel, 2009). In global change and climate change research, vulnerability is an integrative measure of the threats to a system (IPCC, 2001). There are many different definitions of vulnerability, but a common phenomenon across definitions is the link between vulnerability and risk. According to Chambers (1989, p. 1), "vulnerability refers to exposure to contingencies and stress, which is defencelessness, meaning a lack of means to cope without damaging loss." Adger (1999, p. 249) defines vulnerability as: "[t]he exposure of individuals or collective groups to livelihood stress as a result of the impacts of such environmental change." Climate change vulnerability studies combine natural and social science perspectives. Wisner, Blaikie, Cannon and Davis (2004, p. 11) refine the common-sense meaning of vulnerability as being "the characteristics of a person or group and their situation that influence their capacity to anticipate, cope with, resist and recover from the impact of a natural hazard (an extreme natural event or process)." They further explain that vulnerability involves a combination of factors that determine the degree to which someone's life, livelihood, property and other assets are put at risk by a discrete and identifiable event (or series or 'cascade' of such events) in nature and in society.

Thus, vulnerability is a set of prevailing or consequential conditions, which adversely affect the community's ability to prevent, mitigate, prepare for or respond to hazard events. These long-term factors, weaknesses or constraints affect a household's, community's or society's ability (or inability) to absorb losses after disasters and to recover from the damage (Global Crisis Solutions, 2012). The working definition of vulnerability by the IPCC (2007b) p. 883) is "the degree to which a system is susceptible to, and unable to cope with adverse effects to climate change, including climate variability and extremes." The IPCC explains vulnerability to climate change as a function of exposure, sensitivity and adaptive capacity. This can be expressed as follows:

$$V = f(E, S, C)$$

V = Vulnerability
E = Exposure
S = Sensitivity
C = Adaptive Capacity

Thus, vulnerability research have some common elements of interest: the shocks and stresses experienced by the social-ecological system, the response of the system and the capacity for adaptive action (Adger, 2006). Vulnerability is a complex combination of three broad interrelated spheres: physical/material vulnerability, social/organisational vulnerability and motivational/attitudinal vulnerability (Anderson & Woodrow, 1989). These various types of vulnerabilities have interlocking relations to each other in a practical situation (Figure 3.1). Maskrey (1998) further categorises vulnerability as physical, technical, economic, environmental, social, political, cultural, educational and institutional vulnerability. Moreover, vulnerability is a complex combination of interrelationships, mutually reinforcing and dynamic factors that need an inter-disciplinary approach to understand.

3.2.1 *Vulnerability assessment*

Vulnerability assessment is an integral tool of risk assessment. This assessment is the process of estimating the susceptibility of elements at risk (people, household, community facilities and services, economic activities and the natural environment) to various hazards and analysing the root cause which places these elements at risk. The assessment takes into account the physical, geographical, economic, social, political and psychological factors which cause some people to be particularly exposed to the dangers of a given hazard while others are relatively protected (Global Crisis Solutions, 2012).

Vulnerability is a dynamic phenomenon that is often in a continuous state of flux as both the biophysical and social processes that shape local conditions and the ability to cope are themselves dynamic (O'Brien, Eriksen, Lygaard, & Schjolden, 2007). Thus, measurement of vulnerability must therefore reflect social processes, as well as material outcomes within systems. Vulnerability is, therefore, not easily reduced to a single metric nor easily quantifiable (Adger, 2006).

Despite the range of approaches to measuring vulnerability, several best practices have emerged. A basic formula recurrent throughout the vulnerability assessment literature is: Risk + Response = Vulnerability (Moret, 2014). Hoddinott and Quisumbing (2003, p. 46) pose five questions that a vulnerability assessment should answer:

i)   "What is the extent of vulnerability?"
ii) "Who is vulnerable?"

iii) "What are the sources of vulnerability?"
iv) "How do households respond to shocks?" and
v) "What gaps exist between risks and risk management mechanisms?"

Answering these questions requires multiple data collection methods and analytical techniques. Moreover, a final key feature of vulnerability assessment is the inclusion of community perceptions of vulnerability into the assessment design and definition of vulnerability (Kalibala, Schenkb, Weissc, & Elsond, 2012). Vulnerability assessment incorporates a significant range of parameters in building quantitative and qualitative pictures of the processes and outcomes of vulnerability (Adger, 2006). Therefore, the present study used both qualitative and quantitative approaches in assessments of vulnerability to capture the complexity of vulnerability. The following sections (3.2.2–3.2.4) feature the vulnerability assessment approaches which have been used to create a comprehensive baseline for analysis of this research.

3.2.2 The Disaster Crunch Model

A framework that can be useful for understanding vulnerability is the 'Disaster Crunch Model.' This model shows that vulnerability, which is rooted in socio-economic and political processes, has to be addressed first to reduce the risk of disaster (Oxfam GB, 2012). First developed by Blaikie et al. (1994), this model helps us to understand and analyse people's vulnerabilities to disasters. Following the idea of this model, the present study has used a gender-sensitive Disaster Crunch Model (Figure 3.2) which underpins the vulnerabilities of women in a certain socio-economic and cultural context. This means that women experience different levels and types of vulnerability to disasters, including those caused by climate change, in a distinctly different way to men.

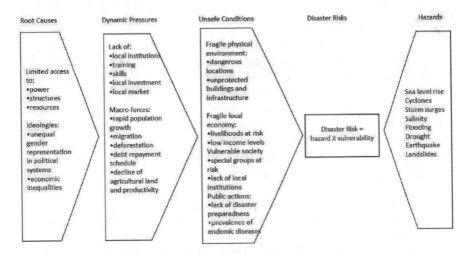

Figure 3.2 Disaster Crunch Model: progression of vulnerability

(Source: modified from Wisner et al., 2004)

The 'Disaster Crunch Model' in Figure 3.2 shows that there are three layers of social processes that cause vulnerability: root causes, dynamic pressures and unsafe conditions. "Unsafe conditions" may be: poor housing conditions, danger-ous location, risky livelihoods, lack of disaster preparedness skills, etc. "Dynamic pressures" may be: no community organisation for collective efforts to reduce hazard risks, lack of training and appropriate skills, rapid migration, lack of local markets and investment for women, etc. "Root causes" may be: lack of cyclone warning systems, women not being allowed to attend meetings on emergency response preparedness, etc. Conversely, "Hazards" may refer to the natural events that may affect different places singularly or in combination (sea level rise, cyclones, water logging, flood, salinity, etc.) at different times (season of the year, time of day, over return periods of different duration). The hazard has varying degrees of intensity and severity. The basic idea of the model is that a disaster risk is the intersection of two opposing forces: those processes generating vulnerabil-ity on one side, and the natural hazard event (or, sometimes, a slowly unfolding natural process) on the other. This model helps researchers to understand people's vulnerability to disasters more fully, and in terms of gender relations, it describes the various levels and types of vulnerability experienced by women in disasters, including those caused by climate change events (Oxfam GB, 2012).

3.2.3 The Livelihood Vulnerability Index (LVI)

Vulnerability assessment and mapping in the context of climate change impact on human systems is a rapidly growing field, and several reviews of the evolu-tion of vulnerability science in relation to environmental hazards have been com-pleted in order to rectify the gaps among climate change vulnerability assessments (Vincent, 2004; McLaughlin & Dietz, 2008). In recent times, the field of climate vulnerability assessment has emerged to address the need to quantify how com-munities will adapt to changing environmental conditions (Hahn et al., 2009). Quantitative measures complement rich narratives of qualitative assessments of vulnerability in places and contexts (Adger, 2006).

Various researchers have contributed to develop methodologies to assess vul-nerability; the indicator approach is one technique that is commonly employed to measure vulnerability to climate change and variability. The indicator approach involves the selection of indicators that a researcher considers to largely account for vulnerability (Deressa, Hassan, & Ringler, 2009). Studies that utilise an indi-cator based index approach are obliged to apply subjective judgement regarding which component and subcomponent to include in the index, and regarding which direction indicators should push the vulnerability index (Sullivan, Meigh, & Fediw, 2002; Vincent, 2004; Sullivan & Meigh, 2005). Therefore, the main chal-lenge of this kind of vulnerability index computation is the selection of major components and subcomponents, and using them in deriving the Livelihood Vul-nerability Index (LVI).

Vincent (2004) and Sullivan et al. (2002) focus on vulnerability to changes in water resources as a result of climate change and combine a number of indicators into various sub-resources that are further aggregated into an overall vulnerability index. Vincent (2004) provides an analysis of national level data primarily from the World Bank to create a Social Vulnerability Index (SVI). Conversely, Sullivan et al. (2002) offers a holistic model for calculating a Water Poverty Index (WPI) that can measure water stress at the household and community levels. The WPI uses a similar structure to that of the Human Development Index (HDI), with the distinct difference that HDI are generally applied nation-wide while the WPI focuses on much more local scales. Based on their original WPI framework, Sullivan and Meigh (2005) also propose a Climate Vulnerability Index (CVI) that can theoretically be applied at the national, sub-national and community levels to assess human vulnerability to changes in water availability as a result of climate change. They suggest a composite index approach with equal weighting to calculate the overall CVI and derivation of the data from various secondary sources. Observed limitations, however, have been found in secondary data analysis and weighting of the composite index. Conversely building on his SVI framework (2004), Vincent (2007) developed two measurements, namely the National Adaptive Capacity Index (NACI) and the Household Adaptive Capacity Index (HACI). These tools are used to assess 'adaptive capacity' rather than 'social vulnerability' of a study population when faced with climate change-induced changes in water availability at the national and household level respectively. The observed limitations of these studies include the calculation of weighted average and non-replicability and non-applicability to future studies.

In the case of using the composite index approach, it must be decided whether or not the components of the index will be weighted and what method will be used if they are not weighted. Eakin and Bojórquez-Tapia (2008) note that equal weighting makes an implicit judgement about the degree of influence of each major component and argue that weighting must be employed in order to most accurately reflect the real world. Hahn (2008) combined techniques and ideas of a variety of climate vulnerability assessments to conduct a livelihood vulnerability assessment in Mozambique. Hahn's study proposes household surveys as the source of data and presents a new framework for grouping and aggregating indicators into the Livelihood Vulnerability Index (LVI) at the district level. The LVI uses multiple indicators to assess exposure to natural disasters and climate variability, social and economic characteristics of households that affect their adaptive capacity, and current health, food and water resources that determine their sensitivity to future climate change impacts.

In light of the strengths and flaws in the previous studies, this study, employs an adaption of Hahn (2008), aiming to assess the livelihood vulnerability of women under extreme climatic events and climate variability by estimating the LVI. While Hahn (2008) used fewer components to construct her LVI, the present study has used a LVI based on livelihood capitals, which offers a further specific categorisation of components. Moreover, the present study focuses on a much broader context of vulnerability by identifying 15 major components under five livelihood capitals, whereas Hahn (2008) and followers used only seven major components to calculate their LVI. By considering five livelihood capitals (i.e. human, natural, financial, social and physical) of the SLF in constructing the LVI, the present study is able to provide a more comprehensive estimation of livelihood vulnerability. It also gives an indication of capital specific vulnerability of women, which is more useful than one based solely on limited components of the LVI.

Overall, this study considered 15 major components altogether under the five capitals classifications. The categorisation of major components in terms of capitals is presented in Table 3.1. Each major component includes several subcomponents that were developed based on available data collected through a household

Table 3.1 Categorisation of major components and subcomponents by five livelihood capitals

Livelihood capital	Major component	Subcomponent
Human	Health	Percentage of women's HHs who have experienced health damage due to natural disasters over 10 years.
		Percentage of women who have limited or no access to medical facilities.
		Percentage of women who have limited or no access to family planning facilities.
		Percentage of women's HHs who have inadequate sanitation facilities.
	Knowledge and skills	Percentage of women who are illiterate or only have primary level education.
		Percentage of women who do not have any training in disaster preparedness.
	Livelihood strategies	Percentage of women who are mainly housewives and have no regular income.
		Percentage of women's HHs dependent on agriculture as major source of income.
		Percentage of women's household heads who work as day labourers.
Natural	Land	Percentage of women's HHs who have no access to cultivable land.
		Percentage of women who have no access to land ownership.
		Percentage of women's HHs who have experienced land degradation due to climate change events.
	Water	Percentage of women's HHs who do not have access to safe drinking water.
		Percentage of women who are the main collectors of water.
		Percentage of women who use pond or river water for household work.
	Natural resources	Percentage of women who use mainly forest resources as cooking fuel.
		Percentage of women who are the main collectors of cooking fuel (firewood).
		Percentage of women's HHs who do not have any trees.

Livelihood capital	Major component	Subcomponent
		Percentage of women's HHs who mainly depend on forest resources for household income.
	Natural disasters and Climate variability	Average number of natural disasters during the last 10 years.
		Percentage of women who were affected by Cyclone Aila.
		Percentage of women who observed climate change consequences in the vicinity.
		Average number of observed consequences due to climate change.
Financial	Annual income	Inverse of income index.
		Percentage of women's HHs who are in break-even positions or in deficit.
	Asset ownership	Percentage of women without a fishing boat and net ownership.
		Percentage of women without livestock ownership.
		Percentage of women without poultry ownership.
	Finance	Percentage of women without adequate access to banking facility.
		Inverse of credit index.
Social	Socio-demographic profile	Average family size of women's HHs.
		Average number of female members per HHs.
		Dependency ratio.
	Social networking	Percentage of women whose family members migrated to other places due to climate change effects.
		Percentage of women who do not have adequate access to local government organisations.
Physical	Housing	Percentage of women's HHs without a solid (*pucca*) house.
		Percentage of women's HHs who have experienced house damage due to Cyclone Aila.
	Household asset	Percentage of women who have no radio or television to access the disaster-related information.
		Percentage of women who have no jewellery (gold/silver) to sell during times of disaster.
	Communication and electricity	Percentage of women without adequate road and transport facilities.
		Percentage of women without electricity facilities.

*HHs = Households

survey and focus group discussion (FGD) in the study area. In total, 41 subcomponents were considered in the present study. The detailed calculation procedure used in the LVI is described later in this chapter under 'data analysis.'

3.2.4 LVI-IPCC: IPCC framework approach

The current study has also applied an alternative vulnerability index called the LVI-IPCC which incorporates the IPCC (2007, p. 883) vulnerability definition: "vulnerability is a function of the character, magnitude and rate of climate change and variation to which a system is exposed, its sensitivity, and its adaptive capacity." In this concept, the components, 'exposure' and 'sensitivity' create potential impacts and increase vulnerability, whilst 'adaptive capacity' decreases it. Therefore, the three main components that need to be considered in LVI-IPCC are exposure, sensitivity and adaptive capacity.

Exposure is "the nature and degree to which a system is exposed to significant climatic variations" (IPCC, 2001, p. 987). Exposure is defined as the degree of climate stress placed upon a particular unit analysis; it may be represented as either long-term changes in climate conditions, or by changes in climate variability, including the magnitude and frequency of extreme events (IPCC, 2001). It is anticipated that exposure to different shocks and stresses, such as a rise in temperature and sea level, cyclones, floods, land erosion and droughts, will be intensified due to climate change (outlined earlier in Chapter 2). Repeated exposure can also result in the loss or destruction of people's resources, adaptive capacity and resilience, leading to greater vulnerability and preventing their quick recovery (Ford, Smit, & Wandel, 2006). Although shocks and stresses are often used interchangeably to denote climate exposure, for this study, extreme events such as cyclones and floods are termed as 'shocks,' whilst the slow onset of phenomena such as the increase in salinity level and rise in temperature are termed as 'stresses.'

The 'sensitivity' of a system to climate change reflects the "degree to which a system is affected, either adversely or beneficially, by climate variability or change" (IPCC, 2007, p. 881). The effect may be direct (e.g., a change in income and health due to different shocks) or indirect (e.g., damages caused by an increase in the frequency of coastal flooding due to sea level rise). Sensitivity reflects the responsiveness of a system to climatic influences and the degree to which changes in climate might affect it in its current form. Thus, a sensitive system is highly responsive to climate and can be significantly affected by even small climate changes.

'Adaptive capacity' refers to the potential or capability of a system to adjust to climate change, including climate variability and extremes, so as to moderate potential damages, to take advantage of opportunities or to cope with consequences (Smit & Pilifosova, 2001). As the name suggests, adaptive capacity is the capability of a system to adapt to the impacts of climate change. Smit, Burton, Klein and Wandel (2000) have identified the following seven factors that determine adaptive capacity which are: wealth, technology, education, institutions, information, infrastructure and social capital. Adaptive capacity can be enhanced by practical means of coping with changes and uncertainties in climate, including variability and extremes (Smit & Pilifosova, 2001). In this sense, enhancement of adaptive capacity reduces vulnerabilities, increases adaptation and promotes sustainable development (Cohen, Demeritt, Robinson, & Rothman, 1998). In functional form, the vulnerability with these three components can be expressed as follows:

$$V = f(\text{E, S, AC})$$

Traditionally, the first two determinants (exposure and sensitivity) have been viewed as dictating the potential for adverse consequences to occur (or vulnerability), thereby providing an indication of potential susceptibility to adverse impacts. Conversely, the third determinant (adaptive capacity) reflects the ability of the system to manage, and thereby reduce vulnerability. Therefore, caution must be exercised to avoid interpreting any of these concepts in an overly rigid fashion (Preston & Stafford-Smith, 2009).

In the context of this study, Table 3.2 shows the organisation of the previously identified 15 major components incorporated into the LVI-IPCC framework. The vulnerability index is constructed on the notion that vulnerability is a function of exposure to climate change and variability, sensitivity to the impacts of that exposure, and the ability to adapt to ongoing and future changes (Hahn, Riederer, & Foster, 2009). This index is used to assess exposure to natural disasters and climate variability; social, economic, and institutional characteristics of households; livelihood strategies that affect their adaptive capacity; and health, natural, physical and financial resource endowments that determine sensitivity to climate change

Table 3.2 Categorisation of major components into IPCC contributing factors to vulnerability

IPCC contributing factors to vulnerability	Major components
Exposure	i. Natural disaster and climate variability
Adaptive capacity	i. Socio-demographic profile
	ii. Social network
	iii. Livelihood strategies
	iv. Asset ownership
	v. Finance
	vi. Knowledge and skills
Sensitivity	i. Health
	ii. Land
	iii. Water
	iv. Natural resources
	v. Housing
	vi. Household assets
	vii. Communication and electricity
	viii. Annual income

impacts. The calculation process of IPCC-LVI is presented later in this chapter under the 'data analysis' section.

3.3 Selection of the study area

As stated previously, millions of people in the coastal areas of Bangladesh are under threat from climate change and climate variability issues. For example, the cyclones that occurred in 1970, 1985, 1991, 1997, 2007 and 2009 caused huge losses and displaced millions of people in the coastal areas (Akter, 2009). Bangladesh is the world's third most vulnerable country to sea level rise in terms of the number of people affected, and among the top 10 countries in terms of the percentage of people living in low-lying coastal zones (Pender, 2008). It is reported that over 35 million people will be displaced from 19 coastal districts of Bangladesh under the prediction of a one-metre sea level rise this century (Rabbani, 2009). The coastal areas of Bangladesh have already been facing salinity problems, which are expected to be further exacerbated by climate change and sea level rise, as sea level rise is causing unusual heights of tidal water. Rising sea level is also causing water level rise in the rivers and thereby accelerating the risks of flooding and water logging (Shamsuddoha & Chowdhury, 2007). Consequently, it can be assumed that the coastal areas of Bangladesh are the most disaster-prone and most vulnerable to the impacts of climate change.

To examine this assumption more clearly, two unions in the Shyamnagar upazila (sub-district) of the coastal district of Satkhira have purposively been selected as the study areas for this study. This upazila was selected as it has been one of the areas most affected by natural disasters and climate change–related impacts in recent times. Shyamnagar upazila is situated in the south-western part of Bangladesh near the Sundarbans, the single largest mangrove patch in the world. The area is located on the periphery of the Bay of Bengal and therefore faces all the challenges associated with climate change. The most visible climate change impacts in this region are natural disasters, such as cyclones and floods, salinity, water logging, increasing temperature and sea level rise. These impacts are outlined further in Chapter 4.

Within this upazila, Gabura and Padmapukur are two of the worst-affected unions (small administrative units) where most of the households have been affected by various climate change effects. These two unions were also affected severely during Cyclone Aila, when most of the people lost their livelihoods (UNDP, 2009). The detailed effects of Cyclone Aila are also presented in Chapter 4. Given the level of impact and the still clearly visible devastating effects of Cyclone Aila, the unions of Gabura and Padmapukur were specifically selected as the study locations for this research. It was assumed that by selecting these two locations which share the same socio-economic characteristics, that a more nuanced understanding of issues on the ground in this region could be gained as they are as representative of similar coastal regions throughout the country (further detailed description of the study area with respective maps is presented in Chapter 4). To measure the climate change impacts on women's livelihoods, the researchers made several visits to these unions over several months and studied the impacts using different survey methods.

3.4 Research design

In order to assess the impact of climate change on the livelihoods of women, the study followed a 'triangulation' process (Flick, 2009) where both qualitative and quantitative studies have been carried out to obtain the research dataset (Figure 3.3). This mixed-method approach allowed for triangulation, which resulted in greater reliability and validation of the data (Marsland, Wilson, Abeyasekera, & Kleih, 2000).

Quantitative research methods attempt to maximise objectivity, replicability and generalisability of findings, and are typically interested in prediction. The essential feature of this approach is that the researcher will set aside his or her experiences, perceptions and biases to ensure objectivity in conducting the research and the conclusions that are drawn (Lincoln & Guba, 1985). On the other hand, qualitative research methods focus on discovering and understanding the experiences, perspectives and thoughts of participants; that is, qualitative research explores meaning, purpose or reality (Hiatt, 1986). The mixing of these two methods into a single study represented an attempt to legitimise the use of multiple approaches, concepts and language in answering research questions, rather than restricting or constraining the researchers' choices (Johnson & Onwuegbuzie, 2004).

Quantitative methods (e.g. the household survey) were used mainly for collecting data on context, whereas qualitative methods (e.g. the focus group discussions, key informant interviews, photographic interpretation and personal observations) were used to obtain rich, detailed and contextually grounded, in-situ data. The two types of data (quantitative and qualitative) provided validation for each other and also created a solid foundation for drawing conclusions about the intervention of climate change on women's livelihoods. The combination of these two approaches was useful in understanding contradictions between qualitative

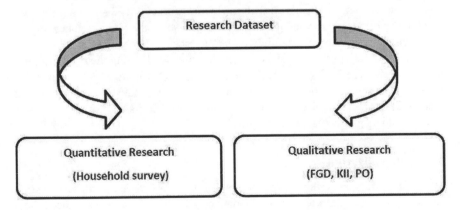

Figure 3.3 Integration of quantitative and qualitative research in research design

and quantitative data, reflecting on participants' views, offering flexibility in the adaption of several research methods, and in collecting rich and comprehensive data sets (Wisdom & Creswell, 2013).

3.4.1 Data sources

The relevant primary data were collected through a household survey targeting women, focus group discussions (FGDs) and key informant interviews (KIIs). Secondary data regarding climate variables and events in the study area, detailed descriptions of the study area and livelihood status information were obtained from various sources such as: the Bangladesh Bureau of Statistics (BBS); Bangladesh National Web Portal; the *Upazila Parishad Manual*; different publications of government; NGOs and international organisations; academic journals and books; and newspaper accounts.

3.4.2 Sampling procedure

The study utilised a multistage sampling technique to obtain a representative sample of those to be interviewed. Firstly, the two unions, Gabura and Padmapukur, were purposively selected considering their close experience of climate change effects, and then a further seven villages from each of the unions were selected for the current study to provide a wider illustration of the effect of impacts. The villages chosen were: Gabura, Khalishabunia, Dumuria, Kholpetua, Khalisha, Chakbara and Jelekhali in Gabura union; and Sonakhali, East Patakhali, Pakhimara, Khutikata, Patakhali, Kamalkathi and Chor Chondipur in Padmapukur union. Having selected the villages, the study then considered what was a reasonable number of samples to be representative of the majority of the population in this region. This, it was assumed, would be around 120; in total 150 women (over 18 years old) were randomly selected and personally interviewed.

In addition, the study conducted six FGDs with 15–20 women and two to three men participating in each group. The male members of the FGDs were usually the community members of that area, such as Union Parishad[1] members, village leaders and NGO workers, who could give a general idea about the climate change effects and statistical information regarding the livelihoods of women in the study area.

Their observations were helpful in explaining and validating the perceptions of women during the FGDs. However, they were not allowed to express their opinion before women or to dominate women's discussion. Finally, 10 key informants who had firsthand knowledge about the community and its institutions were also interviewed as part of the present study.

3.5 Data collection

The household survey through to personal interviews, focus group discussions and key informant interviews were the data collection techniques used throughout this study and are outlined below.

3.5.1 Household surveys through personal interviews

This study conducted 150 face-to-face personal interviews with the women of the sample households to address the research questions. Women were randomly selected without any social, economic or personal bias.

3.5.2 Focus group discussion (FGD)

In total, six FGDs were held with groups of 15–20 women being in each group. The focus groups were formed with women from different age groups. The FGDs were conducted to gather data on climate change experience, livelihood vulnerability, coping strategies and adaptation possibilities related to climate variability and change.

3.5.3 Key informant interviews

Local government and NGO officials who have firsthand knowledge about the women in the study area can give insightful information about their livelihoods, and those who were engaged in relevant policy making and implementation were interviewed for this study.

3.5.4 Direct observation through a transect walk

A number of transect walks were carried out, along with some community people, in both of the unions of the study area. Photographs were taken to capture the actual situation and validate information collected during the field survey.

3.6 Data analysis

In an attempt to bring structure to the data collected for this study, several analyses were conducted to understand the livelihood and vulnerability of women in the coastal areas of Bangladesh. Both quantitative and qualitative analyses have been used to explain the collected data.

3.6.1 Quantitative analysis

Most of the data from the personal interview questionnaires were analysed quantitatively. The livelihood capital status of women based on the utilisation of the SLF has been presented in Chapter 5. The livelihoods of women have been analysed under the five livelihood capitals, that is, human, natural, financial, social and physical. The analysis gives an overall picture of the livelihoods of women. The study also compared the livelihood of women 'before' and 'after' an extreme event (Cyclone Aila) since climate change impacts are not always immediately recognised until prolonged exposure illuminates them. Extreme events such as cyclones, salinity, increased temperatures, heavy rainfall, water logging and sea level rise are all taking place in the study area. For the respondents, however, it was difficult to specifically quantify individual impacts from these events on their livelihood. Therefore, the study chose to consider Cyclone Aila as the base event of climate change, and capital outcomes were compared 'before' and 'after' this climate change-driven natural disaster. Since the short and long-term effects of Aila were extremely evident and people of the study area are still currently facing the adverse consequences, the respondents were able to answer questions on this event more precisely. The analysis identifies specifically how different shocks, trends and seasonality affect livelihood assets and produce the livelihood outcomes of women in the study areas.

The vulnerability contexts are also described in terms of different climate change hazards and the severity of their effects, following the concept of the 'Disaster Crunch Model' (presented in Chapter 6). It identifies and explains the extent of vulnerabilities that developed due to several climate-induced hazards, and consequently enhanced the disaster-associated risks within the community. Moreover, the coping mechanisms which can potentially reduce the level of vulnerability and accessibility of women to various welfare facilities are also discussed in Chapter 6.

In addition to the above-mentioned analyses in regards to understanding livelihood structure and the livelihood vulnerability of women, two vulnerability indexes (LVI and IPCC-LVI) have been used to quantify the livelihood vulnerability using some mathematical computations. The results in Chapter 7 provide us with two index numbers that can subsequently be used to express the level of livelihood vulnerability of women in the study area. The calculation procedures of LVI and IPCC-LVI are described below.

3.6.1.1 Calculation of LVI

The LVI uses a balance weighted average approach (Hahn et al., 2009) where each subcomponent contributes equally to the overall index, even though each major component comprises a different number of subcomponents. The LVI uses the simple method of applying equal weight to all major components. Since each of the subcomponents is measured on a different scale, it is first necessary to standardise each as an index. The equation that is used here for developing the index is by standardisation or normalisation of observed data is similar to the Human Development Index, which is the ratio of the difference of the actual life expectancy and a pre-selected minimum and the range of pre-determined maximum and minimum life expectancy. In the same manner, the standardised index for each subcomponent is determined by the following equation:

$$index_{sc} = \frac{S_v - S_{min}}{S_{max} - S_{min}} \tag{1}$$

where S_v is the value of subcomponent for the sample studied; S_{min} and S_{max} are the minimum and maximum values respectively for that subcomponent. After standardising each subcomponent value, the value of each major component is calculated by the following equation:

$$M_c = \frac{\sum_{i=1}^{n} index_{sci}}{n}$$ (2)

where M_c is the value of the major component; $index_{sci}$ are the indexed subcomponent indicators that make up each major component M_c; and n is the number of subcomponents in each major component. This will produce the value for 15 major components. These values are then used to obtain the five values for livelihood assets [HC (human capital), NC (natural capital), FC (financial capital), SC (social capital) and PC (physical capital)] by the following equation:

$$C_v = \frac{\sum_{i=1}^{n} w_{M_i} M_{ci}}{\sum_{i=1}^{n} w_{M_i}}$$ (3)

where C_v is the value of any capital which equals the weighted average of the major components for this capital. The weight of each major component is represented by w_{M_i} and M_{ci} is the indexed value of major components for the respective capital.

For example, the value of human capital is estimated as:

$$C_{V_H} = \frac{\sum \{(wi * M_{H1}) + (wii * M_{H2}) + (wiii * M_{H3})\}}{wi + wii + wiii}$$ (4)

Where *wi*, *wii* and *wiii* are the number of subcomponents of respective components M_{H1}, M_{H2} and M_{H3}. After obtaining the values for the five livelihood capitals in a similar fashion, the weighted average value of the LVI for women in the study area is obtained by applying the following equation:

$$LVI_w = \frac{\sum_{i=1}^{5} w_{C_i} C_{vi}}{\sum_{i=1}^{5} w_{C_i}}$$ (5)

where LVI_w is the value of the Livelihood Vulnerability Index, equals the weighted average of the five livelihood capitals. The index value of each capital and their respective weight are represented by C_{vi} and w_{C_i} respectively.

This can also be expressed as:

$$LVI_w = \frac{w_H HC + w_N NC + w_F FC + w_S SC + w_P PC}{w_H + w_N + w_F + w_S + w_P}$$ (6)

where LVI_w, the Livelihood Vulnerability Index for women, equals the weighted average of the five livelihood capitals. The LVI is ranged from 0 to 1 where 0 denotes the least vulnerable and 1 the most vulnerable.

3.6.1.2 Calculation of IPCC-LVI

In estimating the IPCC-LVI, the variable that is used under 'exposure' is the index of natural disaster and climate variability (NDCV). 'Sensitivity' comprises the variables of health, land, water, natural resources, housing, household assets, communication and electricity and annual income indices. 'Adaptive capacity' is reflected by the indices of socio-demographic profile, social network, livelihood strategies, asset ownership, finance, and knowledge and skills. Weighted averages of normalised variables were used to form indices for exposure, sensitivity and adaptive capacity using the following equation:

$$CF_w = \frac{\sum_{i=1}^n w_{M_i} M_i}{\sum_{i=1}^n w_{M_i}} \tag{7}$$

where CF_w is an IPCC-defined contributing factor (exposure, sensitivity or adaptive capacity) for women in the study area, M_i are the major components indexed by i, w_{M_i} is the weight of each major component, and n is the number of major components in each contributing factor. Once exposure, sensitivity and adaptive capacity are calculated, the three contributing factors are combined to obtain a vulnerability index using the following equation:

$$LVI - IPCC = (EI - ACI) * SI \tag{8}$$

where the *LVI-IPCC* is the vulnerability index for women expressed using the IPCC vulnerability framework; *EI*, *ACI*, and *SI* are the exposure score for exposure, adaptive capacity and sensitivity respectively. The LVI-IPCC is ranged from −1 (least vulnerable) to +1 (most vulnerable).

3.6.1.3 Vulnerability spider diagram and vulnerability triangle

A vulnerability spider diagram and a vulnerability triangle are used to graphically represent the LVI and IPCC-LVI respectively, because they accurately facilitate visualisation of the vulnerability index score for women. This scale begins at 0 and increases, moving towards the outside edge in 0.20 unit increments for both figures. Therefore, the further the spider or triangle web is from the centre, the higher the vulnerability it indicates for women in the study area. The layout of the vulnerability spider diagram and the vulnerability triangle is presented below in Figure 3.4 and these are used to visualise the livelihood vulnerability of women in Chapter 7.

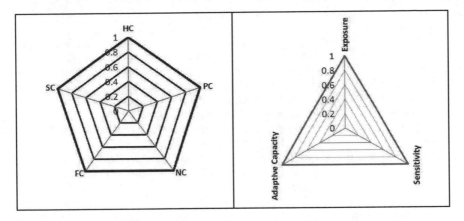

Figure 3.4 Layout of vulnerability spider diagram and vulnerability triangle

3.6.2 Qualitative analysis

The data collected through FGDs and KIIs are qualitative in nature and, therefore, qualitative analysis is used to present the results from these data. The analysis and interpretation of the focus group discussions and key informant interviews need a great deal of judgement and care. The purpose of these interviews was to provide an in depth exploration of climate change impacts on women's livelihood. For analysis of such data, simple descriptive narrative analysis was deemed to be appropriate.

The information gained from the FGDs and KIIs were firstly used to describe the major climate change events and their effects in the study area (Chapter 4).

3.7 Conclusion

This chapter has presented the theoretical framework of this research and has described the research design and methods used in detail. This chapter also highlights the methodological approaches which have been utilised, including sampling, preparation of questionnaires, data collection and data analysis.

Note

1 Union Parishad is the oldest and lowest local government system. At present there are 4550 Union Parishads in Bangladesh. Union Parishads are run by the directly elected representatives (chairman, members, women members). They mainly work for the rural development of the country.

References

Adger, W. N. (1999). Social vulnerability to climate change and extremes in coastal Vietnam. *World Development*, 27(2), 249–269.

Adger, W. N. (2006). Vulnerability. *Global Environmental Change*, 16(3), 268–281.

Akter, T. (2009). *Climate change and flow of environmental displacement in Bangladesh.* Scientific briefing paper series on climate change and development. Unnayan Onneshan-The Innovators, Dhaka.

Anderson, M. B., & Woodrow, P. J. (1989). *Rising from the Ashes: Development strategies in times of disaster.* Boulder and San Francisco: Westview Press; Paris: UNESCO.

Bebbington, A. (1999). Capitals and capabilities: A framework for analyzing peasant viability, rural livelihoods and poverty. *World Development, 27*(12), 2021–2044.

Blaikie, P., Cannon, T., Davis, I., & Wisner, B. (1994). *At risk: Natural hazards, people's vulnerability and disasters.* London and New York: Routledge.

Chambers, R. (1989). Editorial introduction: Vulnerability, coping, and policy. *IDS Bulletin, 20*(2), 1–7.

Chambers, R., & Conway, G. R. (1992). *Sustainable rural livelihoods: Practical concepts for the 21st century* (Discussion Paper No. 296). Brighton, UK: Institute of Development Studies (IDS), University of Sussex.

Cohen, S., Demeritt, D., Robinson, J., & Rothman, D. (1998). Climate change and sustainable development: Towards dialogue. *Global Environmental Change, 8*(4), 341–371.

Deressa, T. T., Hassan, R. M., & Ringler, C. (2009). *Assessing household vulnerability to climate change: The case of farmers in the Nile Basin of Ethiopia* (IFPRI Discussion Paper No. 00935, 18 pp.). Washington, DC: International Food Policy Research Institute.

Devereux, S. (2001). Livelihood insecurity and social protection: A re-emerging issue in rural development. *Development Policy Review, 19*(4), 507–519.

DFID. (1999). *Sustainable livelihoods guidance sheets.* London: Department for International Development (DFID).

DFID. (2000). *Eliminating world poverty: Making globalisation work for the poor* (DFID White Paper). London: Department for International Development.

Eakin, H., & Bojórquez-Tapia, L. A. (2008). Insights into the composition of household vulnerability from multicriteria decision analysis. *Global Environmental Change, 18*(1), 112–127.

Ellis, F. (2000). *Rural livelihoods and diversity in developing countries.* Oxford: Oxford University Press.

Flick, U. (2009). *An introduction to qualitative research.* London: Sage.

Ford, J. D., Smit, B., & Wandel, J. (2006). Vulnerability to climate change in the Arctic: A case study from Arctic Bay, Canada. *Global Environmental Change, 16*(2), 145–160.

Füssel, H. M. (2009). *Review and quantitative analysis of indices of climate change exposure, adaptive capacity, sensitivity, and impacts.* Background note to the world development report 2010. Development and climate change. Potsdam: Potsdam Institute for Climate Impact Research.

Global Crisis Solution. (2012). *Understanding vulnerability: Ensuring appropriate and effective responses.* South Africa: Global Crisis Solutions, Training Unit, Support to Program Quality Project, Nairobi.

Hahn, M. B. (2008). *The livelihood vulnerability index: A pragmatic approach to assessing risks from climate variability and change* (Unpublished Master thesis). Submitted to the Rollins School of Public Health, Emory University, Atlanta.

Hahn, M. B., Riederer, A. M., & Foster, S. O. (2009). The livelihood vulnerability index: A pragmatic approach to assessing risks from climate variability and change—A case study in mozambique. *Global Environmental Change, 19*(1), 74–88.

Hewitt, K. (1983). *Interpretations of calamity: From the viewpoint of human ecology.* Boston, MA: Allen and Unwin.

Hiatt, J. F. (1986). Spirituality, medicine, and healing. *Southern Medical Journal, 79*(6), 736–743.

Hoddinott, J., & Quisumbing, A. (2003). *Methods for microeconometric risk and vulnerability assessments* (Social Protection Discussion Paper). Washington, DC: Social Protection Unit, Human Development Network, World Bank.

Institute of Development Studies (IDS). (2011). *Sustainable livelihoods approaches: Past, present and future?* Brighton, UK: IDS Knowledge Services, with Partners in the Livelihoods Network.

IPCC. (2001). *Climate change 2001: Impacts, adaptation and vulnerability.* Contribution of Working Group II to the fourth assessment report of the Intergovernmental Panel on Climate Change (IPCC) (1032 pp.). Cambridge and New York: Cambridge University Press.

IPCC. (2007). *Climate change 2007: Impacts, adaptation and vulnerability.* Contribution of Working Group II to the fourth assessment report of the Intergovernmental Panel on Climate Change (IPCC) (976 pp.). Cambridge and New York: Cambridge University Press.

Johnson, R. B., & Onwuegbuzie, A. J. (2004). Mixed methods research: A research paradigm whose time has come. *Educational Researcher, 33*(7), 14–26.

Kalibala, S., Schenkb, K., Weissc, D., & Elsond, L. (2012). Examining dimensions of vulnerability among children in Uganda. *Psychology, Health & Medicine, 17*(3), 295–310.

Kollmair, M., & Gamper, St. (2002, September 9–20). *The Sustainable Livelihoods Approach.* Input paper for the integrated training course of NCCR North-South Aeschiried, Switzerland, Development Study Group, University of Zurich (IP6), Zurich.

Lincoln, Y. S., & Guba, E. G. (1985). *Naturalistic inquiry.* Beverly Hills, CA: Sage.

Marsland, N., Wilson, I. M., Abeyasekera, S., & Kleih, U. (2000). *A methodological framework for combining quantitative and qualitative survey methods.* An output from collaborative project between Natural Resources Institute and Statistical Services Centre. Whiteknights: The University of Reading.

Maskrey, A. (1998). *Community based disaster management.* Module on community-based disaster risk management, CBDM-2 hand-out. Bangkok: ADPC.

McLaughlin, P., & Dietz, T. (2008). Structure, agency and environment: Towards an integrated perspective on vulnerability. *Global Environmental Change, 18*(1), 99–111.

Mcleod, R. (2001, May 17–18). *The impact of regulations and procedures on the livelihoods and asset base of the Urban poor: A financial perspective.* Paper presented at the international workshop on regulatory guidelines for urban upgrading, Bourton-on-Dunsmore, Warwickshire.

Moret, W. (2014). *Vulnerability assessment methodologies: A review of the literature.* FHI 360 (89 pp.). USA: United States Agency for International Development (USAID).

O'Brien, K. L., Eriksen, S., Lygaard, L. P., & Schjolden, A. (2007). Why different interpretations of vulnerability matter in climate change discourses. *Climate Policy, 7*(1), 73–88.

Overseas Development Institute. (2003). *Food security and the millennium development goal on hunger in Asia* (Working Paper 231). ODI, London.

Oxfam GB. (2012). *The disaster crunch model: Guidelines for a gendered approach.* Oxford: Oxfam GB.

Pender, J. (2008). Climate change and displacement: Community-led adaptation in Bangladesh. *Forced Migration Review* (31).

Preston, B. L., & Stafford-Smith, M. (2009). *Framing vulnerability and adaptive capacity assessment: Discussion Paper* (CSIRO Climate Adaptation Flagship Working Paper No. 2). Australia: Canberra, ACT.

Rabbani, M. G. (2009). *Climate forced migration: A massive threat to coastal people in Bangladesh*. Clime Asia: Climate Action Network-South Asia (CANSA) Newsletter, Bangladesh Centre for Advanced Studies (BCAS), Dhaka.

Rakodi, C. (2002). A livelihoods approach – Conceptual issues and definitions. In C. Rakodi, & T. Lloyd-Jones (Eds.), *Urban livelihoods. A people-centred approach to reducing poverty* (pp. 3–22). London: Earthscan Publications.

Scoones, I. (1998). *Sustainable rural livelihoods: A framework for analysis* (IDS Working Paper 72). Brighton, UK: Institute of Development Studies.

Shamsuddoha, M., & Chowdhury, R. K. (2007). *Climate change impact and disaster vulnerabilities in the coastal areas of Bangladesh*. Dhaka: COAST Trust.

Smit, B., Burton, I., Klein, R. J. T., & Wandel, J. (2000). An anatomy of adaptation to climate change and variability. *Climatic Change, 45*(1), 223–251.

Smit, B., & Pilifosova, O. (2001). Adaptation to climate change in the context of sustainable development and equity, chapter 18. In J. J. Mccarthy, O. F. Canziani, N. A. Leary, D. J. Dokken, & K. S. White (Eds.), *Climate change 2001: Impacts, adaptation and vulnerability*. Contribution of Working Group II to the third assessment report of the Intergovernmental Panel on Climate Change (IPCC). Cambridge and New York: Cambridge University Press.

Sneddon, C. S. (2000). Sustainability in ecological economics, ecology and livelihoods: A review. *Progress in Human Geography, 24*(4), 521–549.

Sullivan, C., & Meigh, J. (2005). Targeting attention on local vulnerabilities using an integrated index approach: The example of the climate vulnerability index. *Water Science and Technology, 51*(5), 69–78.

Sullivan, C., Meigh, J. R., & Fediw, T. S. (2002). *Derivation and testing of the water poverty iIndex phase 1: Final report*. London: Department for International Development (DFID) and Natural Environment Research Council.

UNDP (2009). *Field visit report on selected Aila affected areas*. United Nations Development Programme (UNDP), Bangladesh.

Vincent, K. (2004). *Creating an index of social vulnerability to climate change for Africa* (Working Paper 56). Norwich, UK: Tyndall Centre for Climate Change Research, University of East Anglia.

Vincent, K. (2007). Uncertainty in adaptive capacity and the importance of scale. *Global Environmental Change, 17*(1), 12–24.

Wisdom, J., & Creswell, J. W. (2013). *Mixed methods: Integrating quantitative and qualitative data collection and analysis while studying patient-centered medical home models* (AHRQ Publication No. 13-0028-EF). Rockville, MD: Agency for Healthcare Research and Quality.

Wisner, B., Blaikie, P., Cannon, T., & Davis, I. (2004). *At risk: Natural hazards, people's vulnerability, and disasters* (2nd ed.). London: Routledge.

4 An overview of the study area

This chapter gives a description of the study area in respect to its geographical and socio-economic settings. This will help the reader to understand the local context and the livelihood settings of women in such areas. The overall objective of this chapter is to demonstrate how and why the women of this area are particularly vulnerable to climate change, and to reinforce the fact that this region is very different from the rest of the country in the way development may be measured.

4.1 Structure of administrative units in Bangladesh

The structure of administrative units in Bangladesh are divided into seven major regions called 'divisions.' Each division is named after the major city within its jurisdiction that also serves as the administrative headquarters of that division. Each division is further split into several districts. There are 64 districts in total in Bangladesh, each further subdivided into upazilas (sub-districts). Bangladesh, at present, has 488 upazilas, which are the second lowest tier of regional administration in Bangladesh. The area within each upazila, except for those in metropolitan areas, is divided into several unions, with each union consisting of multiple villages. Union councils (or Union Parishads or unions) are the smallest rural administrative and local government units in Bangladesh. At present, there are 4550 unions in Bangladesh. Finally, a village is the smallest territorial and social unit for administrative and representative purposes, and there are over 86,000 villages in Bangladesh. In the metropolitan areas, the administration is governed by a city corporation, and there are 11 city corporations in Bangladesh, which are further divided into several municipalities (Bangladesh National Portal, 2015). The administrative units of Bangladesh are presented in Figure 4.1.

4.2 Outline of the study area

The study was conducted in the two unions of Shyamnagar upazila within the Satkhira district of Khulna division, located on the western coastal zone of Bangladesh. Satkhira district is located on the south-west extremity of Bangladesh along the border with West Bengal, India, in between 21°36′ and 22°54′ north latitudes and in between 88°54′ and 89°20′ east longitudes (Figure 4.2). It is approximately

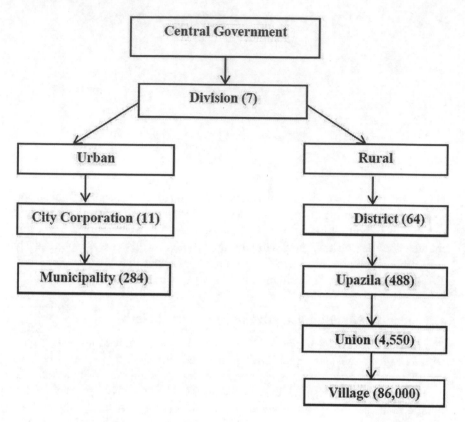

Figure 4.1 Administrative units of Bangladesh
(Source: Bangladesh National Portal, 2015)

3,858 km² in size and the home of two million people. About 1450 km² area of the southern part of this district is occupied by the Sundarbans, the largest single block of tidal halophytic mangrove forest in the world. The population density of Satkhira is 539 per sq. km, and the male-female ratio is 100:106 (Bangladesh National Portal, 2015).

Satkhira consists of seven upazilas and of the seven upazilas of the district, Shyamnagar is the largest. Figure 4.2 shows the map of Satkhira district indicating where Shyamnagar upazila is located. Regionally, the south-west coastal belt of Bangladesh is an intricate system of biodiversity, which includes the Sundarbans. The coastal zone spans over 580 kilometres of coastline and is prone to multiple hazards. Cyclones, floods, tidal surges, high level of salinity, land erosion and periodic water logging are common throughout this region, significantly shaping the lives and livelihood of local communities (Solidaritiés International, 2013). Agriculture and shrimp farming are the major areas of employment and livelihood in the south-west coastal districts. Around 85 per cent of the people in the region

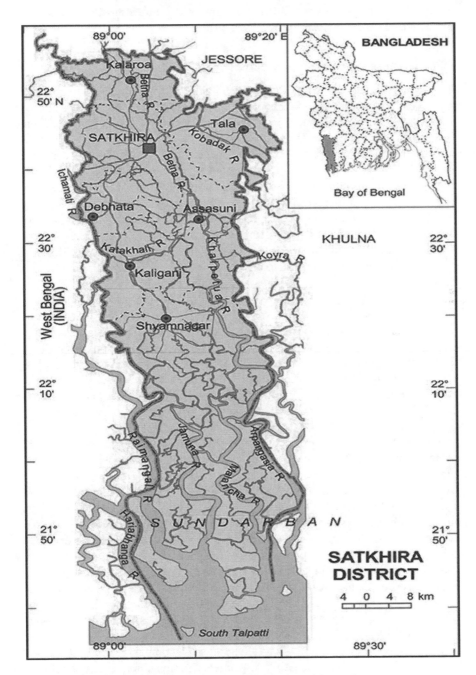

Figure 4.2 Map of Satkhira district showing the location of the study area
(Source: Banglapedia, 2015)

are employed in agriculture and landless farmers make up almost 66 per cent of the population (Abedin, Habiba, & Shaw, 2012).

4.3 Shyamnagar upazila at a glance

Shyamnagar is the largest upazila of the Satkhira district in respect to area and the second largest in respect to population. Bangladesh can be divided into 24 sub-regions based on physical features. Shyamnagar upazila is located within the immature delta region. The main distributaries of the Ganges no longer flow through this area and therefore continuous formation of this estuary is incomplete (MoDMR, 2014). The upazila is the most southern upazila of the Satkhira district and is bounded in the north by the Kaliganj and Assasuni upazilas, in the east by the Koyra upazila of the Khulna district, in the south by the Sundarbans and Bay of Bangal, and in the west by India (Figure 4.2). The upazila occupies an area of 1968 sq. km of which only 483 sq. km are mainland and the rest of the 1485 sq. km belong to the Sundarbans forest area (Bangladesh National Portal, 2015).

The population of Shyamnagar is about 318,254, of which 52 per cent are female and the average household size for the upazila is 4.39 persons (BBS, 2012). More than two-thirds of the population (67 per cent) in this upazila are poor and consume 2122 kcals/person/day or less. More than half (55 per cent) are extremely poor and food consumption is insecure at 1805 kcal/person/day or less whereas in the western countries the average food consumption is about 3500 kcal/person/day (UN, 2010). The literacy rate, especially for women, is low compared to other upazilas of the country. The female literacy rate of this upazila is 43 per cent, whilst the national literacy rate for female is 50.2 per cent (BBS, 2012). The school attendance rate for females is only 51 per cent, compared to the 81 per cent attendance rate for females nationally (BBS, 2012).

The upazila consists of 12 unions and the Gabura and Padmapukur unions selected for this study are located here. Being located on lands which are below the mean sea level, these areas are highly susceptible to coastal hazards like cyclones, storm surges, high tides, flooding, high salinity and land erosion. Apart from these climate-related stressors, these areas are also vulnerable to various non-climatic stressors that include conversion of over 80 per cent of croplands to saltwater shrimp *ghers*,[1] which started in the late 1980s. This practice excluded a huge numbers of landless poor who used to subsist as agricultural labourers in the pre-shrimp regime. Since the areas are adjacent to the Sundarbans, the majority of these people have now become dependent on resource collection from the Sundarbans Reserved Forest (SRF) as an easy alternative, but this in itself has created increased pressures on the SRF and the consequent degradation of the forest (Mangroves for the Future, 2014).

Gabura union consists of 15 villages. The area and population size of this union are 33 and 38,825 square kilometres respectively. In this union, only two kilometres of roads are paved whereas 70 kilometres of roads are mud or dirt (Bangladesh National Portal, 2015). Water logging, scarcity of pure drinking water and poor sanitation are three of the many problems in these villages. Only 30 per cent of households have access to safe drinking water and only 22 per cent of people have access to proper sanitation systems (MoDMR, 2014). The distance of the Upazila Health Complex from this union is 25 kilometres. The main livelihood activities are shrimp farming, fishing and forest resource collection. The soil is not suitable for any crop or vegetable production because of the high levels of salinity (Nowabenki Gonomukhi Foundation, 2013).

Similar to Gabura union, Padmapukur also consists of 15 villages. The area and population size of this union are 35 and 26,447 sq. km respectively. In this union, only five kilometres of roads are paved, whereas 55 kilometres of roads are mud or dirt (Bangladesh National Portal, 2015). Salinity, water logging, scarcity of pure drinking water and arsenic contamination from groundwater are the common problems of this union. Only 40 per cent of households have access to proper sanitation facilities (MoDMR, 2014). The distance of the Upazila Health Complex from this union is 13 kilometres. Similar to Gabura, most of the households of this union are engaged in shrimp farming, fishing and forest resource collection. Other than these activities, some rice farming is also practised in the high lands and midlands that are located upstream. Production of other crops and vegetables is negligible (Nowabenki Gonomukhi Foundation, 2013).

4.4 The Sundarbans: a part of livelihood of the community in the study area

The Sundarbans is the world's largest contiguous single-tract mangrove forest shared between two neighbouring countries, Bangladesh and India. The Sundarbans Reserve Forest (SRF) is located in the south-west of Bangladesh, between the river Baleswar in the East and the Harinbanga in the West, and adjoining the Bay of Bengal. Lying between latitudes 21° 27' 30" and 22° 30' 00" north and longitudes 89° 02' 00" and 90° 00' 00" east, and with a total area of 10,000 km^2, 60 per cent of the Sundarbans lies in Bangladesh and the rest in India. The land area, including exposed sandbars, occupies 414,259 ha (70 per cent) with water bodies covering 187,413 ha (30 per cent) (UNESCO, 2015).

The Sundarbans Reserve Forest (SRF) is located within the administrative districts of Khulna, Bagerhat and Satkhira. It is recognised as a site of national and international importance for conserving biodiversity due to its richness. Three wildlife sanctuaries in the Sundarbans, covering an area of 139,700 ha, have been declared as world heritage sites (798) by UNESCO in 1997, and the entire Sundarbans was declared as the 560th Ramsar Site in 1992. The Sundarbans provides a habitat for large number of plant and wildlife species, including several rare and endangered species of the world. It also provides a home for the largest population of highly endangered Royal Bengal tigers. The unique biota of the Sundarbans is comprised of 334 plant species, 375 faunal species, 49 mammalian species, as many as 400 fish species, 315 bird species and 53 species of reptiles (Sadik & Rahman, 2009). The forest also acts as shelter to protect human habitats from cyclones, tidal surges and saline water intrusion (Biswas, Choudhury, Nishat, & Rahman, 2007).

The SRF provides a buffer for the lives, livelihood and assets of the 3.5 million people who live in its immediate vicinity. They are socio-economically vulnerable groups due to their low income and unstable livelihood opportunities (Kabir & Hossain, 2008). The mangrove forest of the Sundarbans is valuable because of its rich biodiversity, which are commercially exploited, particularly for its Non-Timber Forest Produce (NTFPs), and this epitomises the livelihood of many forest fringe dwellers (Singh, Bhattacharya, Vyas, & Roy, 2010). The main NTFPs collected from the mangrove forest of the Sundarbans include: tannin bark (most Sundarban species like *Ceriops decandra*, *Ceriops myrobalans*, *Phoenix paludosa* yield around 30–42 per cent tannin); *Nypa fruticans* (*golpata*),[2] natural honey from *Apis dorsata*, cultured (apiary) honey (*Apis indica*) and bee wax; fuelwood and small poles and boles; and fishes, prawns, crabs and shrimps; and lime (manufactured from local crustaceans *jorgran*, *kastura* and *jhinuk*). The traditional resource users of the Sundarbans are the local Bawali (woodcutters), the Mouali (honey collectors), the *golpata* collectors and the Jele (fisherman) communities (Hossain & Kabir, 2006).

The people of Padmapukur and Gabura unions largely depend on the natural resources provided by the Sundarbans for their livelihood because they live in the closest and most adjacent locality to the Sundarbans. The ecosystem services of the Sundarbans provide an opportunity for the households of this community to continue their livelihoods while they continuously face the severe challenges of climate change. Therefore, it should be noted here that the Sundarbans and its resources have a significant influence on the livelihoods of households in the study area and this is also reflected in the current research.

4.5 The severe impacts of Cyclone Aila

Cyclone Aila hit the south-western part of Bangladesh on 25 May 2009 and affected an estimated 3.90 million people in 11 coastal districts. The impact was aggravated as the cyclone hit Bangladesh during the high tide cycle, which resulted in tidal surges of up to 22 feet. The two worst-affected districts were Satkhira and Khulna.

The surge of water caused portions of the embankments to collapse and people who believed that the embankments could protect them did not have enough time to evacuate to higher and/or safer ground. A total death toll of 190 persons was recorded. In the aftermath of the cyclone and tidal surges, some 100,000 livestock were killed and over 340,660 acres of cropland was destroyed. The government of Bangladesh also reported that over 6000 kilometres of roads were damaged or totally destroyed, and around 1400 kilometres of flood protection embankments were washed away (MoFDM, 2009). In total, Cyclone Aila made 375,000 people homeless, with many of them seeking refuge on elevated roads and embankments while others were able to seek shelter in schools and other public buildings (International Agencies, 2009). Cyclone Aila fully destroyed 445 and partially destroyed 4588 education facilities/institutions across all of the affected districts, impacting approximately 500,000 children (UN, 2010).

A summary of the effects of Aila on livelihood have been presented in Table 4.1. The main livelihood in the affected areas is fishing, with more than 60 per cent of people directly or indirectly involved in the fishing sectors. About 52,961 acres of shrimp *ghers*, as well as 1074 acres of sweet water fishponds, were damaged by Cyclone Aila. The estimated loss was approximately BDT[3] 1.5 billion. Day labourers and small traders involved in collecting shrimp from farms and on-selling it to mainland wholesalers were also seriously affected. Cyclone Aila completely damaged 3412 acres (46 per cent) of standing crops out of 7392 acres. At the time of Aila, the major standing crops were jute and dry season vegetables. Rice production was very limited in the areas because most cultivable land for rice was being used for shrimp cultivation. Thus, the Sundarbans represented one of the largest sources of livelihood for the people of the Aila-affected area. Many lost most of their tools and equipment used in forest resource collection and many have fallen into debt due to the scarcity of work opportunities post Aila.

In affected areas, traditional houses were constructed largely from earthen walls, with wood to stabilise the structure, and roofs of leaves from the *golpata* plant, collected from the adjacent Sunderbans. These houses immediately collapsed and were washed away following the inundation that was caused by the collapse of the embankments.

According to the loss and damage statements prepared by the District Administration in July 2009, about 76 per cent households in the four upazilas (Shyamnagar,

Table 4.1 Effects of Cyclone Aila on different livelihood groups

Livelihood group	Sub-group	Loss of livelihood assets by Cyclone Aila
Fishing (60 per cent)	Fishing (open water bodies), shrimp fry collector, crab collector (in open water/ Sundarbans), crab culture, shrimp farmer, fish culture (sweetwater pond), Small traders (fish/shrimp/crab, fish and crab feed), Shrimp farming	Fish and shrimp in pond and *gher* washed away, nets, boats, accessories, wages, capital, stock goods, fish feed
Farming (13 per cent)	Small agricultural producer (veg/fruits), poultry farming, farming labour, HH poultry, livestock is rarer	Standing crops, poultry in stock, poultry feed, wage, spade, poultry, goats (90 per cent), cows (20 per cent)
Forest dependent (15 per cent)	Honey collector, leaves (*golpata*) collector, forest labourer	Boat, drams, ropes, axes, tools, wages
Others (12 per cent)	Boat carpenter, net maker, small traders (net), housing labour, carpenter, mason, seasonal migrated labour, rented motor cycle driver, small traders (shopkeeper), service holder	Tools, threads, capital, wages Shop, in stock

Source: International Agencies, 2009.

Asasuni, Dacope and Koira) was affected by Cyclone Aila. Approximately 78 per cent and 73 per cent of families took shelter on the embankments in Gabura and Padmapukur unions respectively during Aila (UN, 2010). Prior to Cyclone Aila, high saline levels were found in the groundwater of these areas. Therefore, people were dependent mostly on surface water bodies. Cyclone Aila resulted in saline intrusion over the surface of water resources, making them unfit for drinking. The Water, Sanitation and Hygiene (WASH) sector assessment of UNICEF carried out during May and June 2009 found that some 4000 protected ponds, 1000 pond-sand filters and 13,000 tube-wells were damaged by *Aila* in the above mentioned four upazilas. In addition, over 210,000 household latrines were fully or partially damaged in those upazilas (UN, 2010).

In Shyamnagar upazila, Gabura, Padmapukur and Burigoalini unions were the most affected by Cyclone Aila, of which Gabura and Padmapukur were severely affected. Almost 100 per cent of the dwelling houses in these two unions were damaged during Aila (UNDP, 2009). In Shyamnagar, in particular, the number of the population and families affected by Cyclone Aila were 158,622 and 33,740 respectively. Cyclone *Aila* also completely damaged 13,223 hectares of shrimp *ghers* and 498 hectares of standing crops. The numbers of poultry and livestock destroyed by this Cyclone were 23,275 and 634 respectively. There were also 966 deep tube-wells and shallow tube-wells destroyed, 158 ponds infiltrated by salt and 2006 sweetwater ponds salinised. In addition, 20,850 toilets were damaged during that time in Shyamnagar upazila.

The figures above reveal that Cyclone Aila destroyed almost all the livelihood assets, such as houses, infrastructures, crops, livestock, fisheries, water and sanitation facilities in the study areas, and this caused prolonged suffering for local people. The present study has considered this disaster as a trigger point because, it significantly changed the livelihoods of women in the study area in many ways and women can remember the 'before' and 'after' Aila situation very clearly since it was the most recent devastating event in that area. The effects of Aila are specifically outlined in Chapter 5.

4.6 Observations of the study area by the researchers

This section presents an overall picture of the study area based on observations and a brief discussion of the logistics needed to reach the study area. As previously mentioned, the study area is located on the edge of Sundarbans and is thus the last locality before the forest. Photo 4.1 presents a look at the Sundarbans in the area that is being studied.

This area is one of the most disadvantaged areas in the country as a whole. To reach the study area, several modes of transportation along the journey were needed (bus, motorcycle, boat and walking). From the capital, Dhaka, it takes 9–10 hours by bus to reach the Satkhira district. Though the distance from Dhaka to Satkhira is only 243 kilometres, the poor level of transportation, condition of the roads and uncertain ferry services make the journey long and tiring. Moreover, the travel time depends upon the availability of the ferry at the Aricha ferry

Photo 4.1 The Sundarbans, adjacent to the study area

terminal. There is no other way to get to Satkhira except by using this ferry ser-
vice on the Padma River. It takes another two hours to reach the upazila sadar[4]
(Shyamnagar) from the Satkhira district, and then another 2–3 hours to reach the
Padmapukur and Gabura unions from Shyamnagar upazila sadar.

On the way from the upazila sadar to those two unions, most of the roads are
gravel roads on which people usually travel by bus or motorcycle. Normally, only
two bus services run on this road per day. Alternatively, rented motorbikes are a
very popular mode of transportation in the area. After a certain distance though, villagers need to take a boat to cross the Kholpetua River to reach the Padmapukur and
Gabura unions. People also need to use boats all year round, because of the many
small rivers and canals of that area which are subject to flooding events. Engine-
driven small boats are also popular forms of transportation to cross the rivers.

The inter-village communications mainly operate via mud roads. In the rainy season it is almost impossible to reach this area. Sometimes, villagers need to walk in up to one foot of deep mud on rainy days (see Photo 4.2). People of these areas are always struggling to cope with this situation. Due to the type of soil and presence of salinity in the water, the soil usually turns very muddy when it rains. Conversely, when there is no rain it is very dry and it turns into strongly bonded solid, brick-like soil. During that time, the roads become very hard, like paved roads, and motorbikes can be used on them (see Photo 4.3a). Usually rickshaws and vans are not seen in these areas, unlike other rural areas in Bangladesh.

Communications in these areas are also a big challenge, as in some villages there are no roads at all. According to the focus group discussants (FGD) (outlined further in Chapter 5), during Cyclone Aila all the existing roads in these villages were destroyed. They also reported that, after Cyclone Aila in 2009, some government and non-government organisations, with the help of international organisations, tried to rebuild some of muddy roads and some wooden bridges (see Photo 4.3b) for villagers in some places, though those initiatives were not sustained for those in undeveloped rural areas.

The overwhelming majority of the coastal residents are poor and live in structurally weak houses (BBS, 2011). It was also seen in the study area that most of the houses are made out of mud and *golpata* (*Nypa fruticans*) roofs. The housing conditions are really too poor to live in, but there are no alternatives for those people (see Photo 4.4). Other infrastructural developments were rarely found in the study

Photo 4.2 Scenarios of study area's mud road in the rainy season

Photo 4.3 Motorbikes in the village's unpaved road (a) and wooden bridge to connect roads in the village (b)

Photo 4.4 Typical houses of the study area

area. There was no electricity in those areas, but there were some solar panels on a few houses, which were provided by some NGOs at a subsidised price. Only four cyclone centres were found in Padmapukur union and none in Gabura union at all. During times of the disaster, only the villagers who live near the centres are able to use the facilities, whilst most of the villagers have no scope to use them.

During the field visit it was seen that the land cover was mostly occupied by shrimp *ghers*. Throughout the entire area, with the exception of the homestead areas, most of the lands had been converted to shrimp *ghers* (see Photo 4.5). It was also evident that there were only a few trees in the whole vicinity. Moreover, there was no livestock seen in the study area, except some experimental goat rearing in

Photo 4.5 Shrimp *ghers* in the study area

some households supported by NGOs. These are unlikely scenarios compared to other rural areas of Bangladesh.

The setting of the villages demonstrates that the economic condition of the people is very poor. Due to the presence of salinity in the soil, people have limited scope to cultivate the land for agricultural purposes. Most of the agricultural lands are used for shrimp farming. It is common to see people, including women and children, catch shrimp fingerlings from the rivers and canals (Photo 4.6).

Photo 4.6 Catching of shrimp fingerlings in riverbank

4.7 Conclusion

It became evident from the field visits to these two unions that they faced several challenges that had been exacerbated by extreme weather events. Cyclone Aila, for example, had inflicted devastating effects on these areas and these were very visible, even four years after the event. Reports from several newspapers on the anniversary of Aila in 2015 supported this observation, stating that people of the Aila-affected areas were still living miserable lives and could not recover from the losses they had incurred due to Aila and other climate-induced problems (*The Daily Prothom Alo*, 26 May 2015). The other climate change effects, such as salinity and water logging, were also prominently visible there. The following chapters examine the livelihoods of women in the study areas, as told by those who lived through Alia and continue to negotiate the ongoing challenges of life in a climate-susceptible setting.

Notes

1 'Gher' is a Bengali word used to describe coastal fisheries in the south-western region of Bangladesh. *Gher* means encirclement of brackish water areas along the coastal belts by building dwarf earthen dykes in order to hold tidal water containing shrimp fries until they grow to marketable size.
2 Commonly known as the nipa palm (or simply nipa) or mangrove palm, is a species of palm native to the coastlines and estuarine habitats of the Bay of Bengal.
3 Bangladeshi Taka: 1 US$ = 78 BDT.
4 Upazila headquarters/center (sub-district headquarters).

References

Abedin, M. A., Habiba, U., & Shaw, R. (2012). Chapter 10, Health: Impacts of salinity, arsenic and drought in South-Western Bangladesh. In S. Rajib, & T. Phong (Eds.), *Environment disaster linkages*. Community, environment and disaster risk management (Vol. 9, pp. 165–193). Bingley, UK: Emerald Group Publishing Limited.

BBS (2011). *Bangladesh – household income and expenditure survey 2010*. Bangladesh Bureau of Statistics, Statistics and Informatics Division, Ministry of Planning, Government of the People's Republic of Bangladesh, Dhaka.

BBS. (2012). *Community report-Satkhira Zila, population and housing census 2011*. Dhaka: Bangladesh Bureau of Statistics, Statistics and Informatics Division, Ministry of Planning, Government of the People's Republic of Bangladesh.

Bangladesh National Portal. (2015). *Bangladesh national portal-information and service in a single window*. Retrieved from bangladesh.gov.bd.

Biswas, S. R., Choudhury, J. K., Nishat, A., & Rahman, M. (2007). Do invasive plants threaten the sundarbans Mangrove forest of Bangladesh? *Forest Ecology and Management, 245*(1–3), 1–9.

Hossain, J. & Kabir, D.M.H. (2006) *Sundarban reserve forest-an account of people's livelihood & biodiversity*. Unpublished Report, Unnayan Onneshan, Dhaka.

Kabir, D. M. H., & Hossain, J. (2008). *Resuscitating the sundarbans: Customary use of biodiversity and traditional cultural practices in Bangladesh* (1st ed., 98 pp.). Dhaka: Unnayan Onneshan-The Innovators.

Mangroves for the Future. (2014). *Enhancing adaptive capacity of Sundarbans dependent community through climate resilient livelihoods.* A project by Mangroves for the Future (MFF) with implementing partner Center for Natural Resource Studies (CNRS).

MoDMR. (2014). *Shyamnagar Upazila risk atlas.* Dhaka: Comprehensive Disaster Management Programme (CDMP II), Ministry of Disaster Management and Relief (MoDMR), Government of the People's Republic of Bangladesh.

MoFDM (2009). Ministry of Food and Disaster Management data, cited in "In-depth Recovery Needs Assessment of Cyclone Aila Affected Areas" conducted by International Agencies (ActionAid, Concern WorldWide, DanChurchAid, MuslimAid, Islamic Relief, Oxfam-GB and Save the Children-UK).

Nowabenki Gonomukhi Foundation (NGF). (2013). *Shyamnagar Upazila information sheet.* Shyamnagar, Satkhira: NGF-A Non Government, Non Profit, Non Political, Voluntary Development Organization.

Sadik, S. & Rahman, R. (2009). *Indicator framework for assessing livelihood resilience to climate change for vulnerable communities dependent on Sundarban mangrove system.* Paper presented in Fourth South Asia Water Research Conference on Interfacing Poverty, Livelihood and Climate Change in Water Resources Development: Lessons in South Asia, May 4-6, 2009, Kathmandu, Nepal.

Singh, A., Bhattacharya, P., Vyas, P., & Roy, S. (2010). Contribution of NTFPs in the livelihood of mangrove forest dwellers of Sundarban. *Journal of Human Ecology, 29*(3), 191–200.

Solidar1tiés International. (2013). *Chronic poverty in the southwest coastal belt of Bangladesh* (A Report). Dhaka: Solidaritiés International and Uttaran.

The Daily Prothom Alo. (2015, May 26). "আইলার ছয় বছর: পানি আছে পানি নেই"-Ailar Choy Bochor: Pani Ache, Pani Nei (Trans.: Six year of Aila: Water is everywhere but there is no usable water). News published in *The Daily Prothom Alo* (Bengali News Paper), Dhaka.

UN. (2010). *Cyclone Aila.* Joint UN multi-sector assessment and response framework. New York: United Nations.

UNDP. (2009). *Field visit report on selected Aila affected areas.* Bangladesh: United Nations Development Programme (UNDP).

UNESCO. (2015). *Sundarbans.* World Heritage Convention, United Nations Educational, Scientific and Cultural Organization (UNESCO). Retrieved from http://whc.unesco.org/en/list/798

5   Livelihood capitals of women against a background of disaster

This chapter analyses women's livelihood capitals on the basis of the Sustainable Livelihood Framework (SLF). The SLF approach is concerned first and foremost with people, and pays particular attention to a range of capitals or assets. Livelihood capitals, together with the external context, result in different livelihood strategies and associated outcomes (Scoones, 1998). As discussed in Chapter 3, capitals are considered to play an important role in enabling a household to generate a means of survival (Ellis, 2000).

The endowments of the five livelihood capitals, that is, human, natural, financial, social and physical capital that directly help to construct livelihoods, are assessed in this chapter in order to understand women's livelihoods in a disaster-susceptible setting. The livelihood capitals are examined from the perspectives of the respondent women in the study and their households, as they are very closely related to each other. Cyclone Aila in 2009 was considered the trigger from which to measure how a climate-induced disaster has changed the livelihood settings of women in the study area. Based on field data collection through personal interviews with women, focus group discussions and key informant interviews, the following sub-sections of this chapter present a detailed assessment of the five livelihood capitals of women in the study area.

5.1 Human capital

Human capital represents the skills, knowledge, ability to labour and good health that together enable people to pursue different livelihood strategies and achieve their livelihood objectives (Kollmair & Gamper, 2002). Human capital can be considered a building block or means to achieve desired livelihood outcomes. In fact, when properly accumulated, it can be regarded as an end in itself. It encompasses not only the quantity of the physical ability embodied in human capital, but also the quality of the human capital (such as skills and knowledge) that enables people to take advantage of economic opportunity (Rakodi, 1999). Human capital is particularly important at a household level, which contributes to the productivity of labour and a person's capacity to manage a household. In the context of the present study, human capital includes: age composition; marital status; the education status of women; their occupation; occupation of the household head; and the daily workloads, each of which are described in detail below.

5.1.1 Age structure

Knowledge of the age structure of sample respondents is important in understanding the potential of productive human resources. The basic criterion for selecting the sample women was that they should be considered an 'adult' in terms of the workforce (18–60 years old) and, as such, would be familiar with several livelihood strategies in the home and workplace. Most of the respondents were born and raised in the study area. The age of the women was classified into four categories, which were: 18–25 years, 26–35 years, 36–50 years and above 50 years. Most of the women of the sample came from the 36–50 years category, which accounted for 41 per cent of the total respondents. If we consider the women from the two age groups incorporating those of 26–50 years of age, 74 per cent of women belonged to this category. This figure represented most of the respondents who were from the relatively more productive middle-aged group. They play a vital role in constructing livelihood strategies. Moreover, because of their age structure, they can easily recall the past climate change events and their related consequences, which is a major component this study seeks to address.

5.1.2 Educational status

Education is a prerequisite for achieving sustainable human resource development. It plays an important role in enhancing productivity at individual and community levels (UNESCO, 1997). Educational attainments help to develop conceptual skills and facilitate the acquisition of technical skills that can strengthen the livelihood of any group. Lower levels of education, thereby, reduce the ability of women and girls to access information, including early warning mechanisms and resources, or to make their voices heard. This poses an extra challenge when women want to look for their livelihood strategies (Rahman, 2013). Therefore, education is considered as a significant component of livelihood.

The educational level of the respondents was classified into four categories. These were: illiterate, can sign only, primary level and above primary level. The results revealed that 34 per cent of the respondents of the study were illiterate and that 17 per cent could sign only. The category 'sign only' meant women were not able to read or write. Many had just learnt how to sign their name because of the influence of local NGOs. Among the remaining respondents, 22 per cent of them had attended primary school, which was up to class 5, and 27 per cent of the respondents had above primary-level education, though most of them had not passed the Secondary School Certificate (SSC) examination. Most of the respondents in this category who had started their secondary education failed to finish due to the distance of the school from their locality and other socio-economic factors, such as early marriage, family financial problems and so on. A respondent named Rafeja Khatun from the Chakbara village of Gabura union said that:

> In our time there was no school in this area. Moreover, our guardians didn't permit us to go to school. But situation has changed. Now girls are going to school and getting various facilities from government.

According to the official definition of 'literacy rate' given by UNICEF and UNDP, this primarily refers to the 15+ age group of people who can read and write. Thus, the literacy rate in the present study was calculated at 49 per cent, which shows that the educational levels of the women in the study areas were low compared to the national and respective divisional levels. In the latest literacy assessment survey of the Bangladesh government, it is documented that the national female literacy rate is 50.2 per cent and the female literacy rate in Khulna division where Shyamnagar upazila is situated is 52.8 per cent (BBS, 2013b). Therefore, it can be said that the women in the study area are still behind the national and divisional status in terms of the literacy rate overall, though obviously this is improving slowly over time.

5.1.3 Marital status

Marriage in Bangladesh is considered to be a social necessity and responsibility and, therefore, helps in explaining the livelihood status of women. Women separated from their husbands or who are widows usually have a low status in the society. Conversely, married women play a significant role in constructing livelihood strategies to achieve their livelihood outcomes. It is common in Bangladesh that most of the middle-aged women are married. Confirming this view, the study also found that about 90 per cent of the respondents were married. Among the remaining respondents, 9 per cent were widowed or divorced, and only 1 per cent were single.

The married women had various responsibilities to their families and, hence, they struggled more to cope with any adverse conditions, including climatic variability and extremes. In most of the cases, they ignored their own wellbeing for the sake of other family members and, therefore, may be recognised as the most disadvantaged group of that society.

5.1.4 Occupational status

The patriarchal nature of Bangladeshi society means that men dominate the formal occupation sector in Bangladesh and often resist the idea of women entering the paid labour force, presumably because of expectations that women will assume the full burden of housework and child care (Efroymson, Biswas, & Ruma, 2007). At a national level, 75 per cent of women did not undertake work that registered as an economic activity, and the figure is slightly higher in the case of rural women (BBS, 2013c). It is considered that the most responsible job for women is to take care of the whole family and to continue to supply food for the family (Asaduzzaman, Reazuddin, & Ahmed, 1997). A recent study has revealed that 71 per cent of rural women in Bangladesh were not interested in paid work. The main reason for this, as stated by the women, was that their families did not want them to be engaged in such work. Moreover, 40 per cent of paid women needed to consult with their families before spending the money they earned. Women in low-income households, however, were still involved in economic activities, mostly around homestead-based production, which contributes up to 16 per cent of the household income in Bangladesh (CPD & MJF, 2014).

The present study also found that 85 per cent of the respondents were mainly housewives and were responsible for maintaining their families. Because of the disadvantaged socio-economic condition of the study area, this figure is higher than the national average. Among the respondents, only a small percentage of women were engaged in income-generating activities outside the home, such as being day labourers (4 per cent) and fry collectors[1] (4 per cent). They were basically poor women who had no productive male persons in their household to work. They had to do such jobs to support their families, although they were not generally appreciated by the society, which remains steeped in patriarchal notions of womanhood. Some women, though, have learnt to look beyond the cultural norms, with a day labourer, Azimon Begum from the Pakhimara village of Padmapukur union, noting that:

> Why I will care about society? When there is no food in my house, nobody will give me any food. I am responsible for managing food for my family members. So, there is no alternative to work outside.

Generally, women in the study area preferred to do some income-generating activities that could be done within their homestead area[2] rather than outside, including things like sewing, which was practised by 3 per cent of the respondents. Among others, 3 per cent of the respondents were too old to work due to being in an older age bracket, but they were a very good source of information about climate variability generally, and 1 per cent of the respondents were full-time students.

5.1.5 Occupational status of the household head

More than 90 per cent of the respondents said that the head of their household was male. That is, men, whether by earning an income or simply being male, are the *de facto* household heads. The occupation of the household heads of the respondents largely determined the women's livelihood standards since they were mostly dependent on them. The occupational structure is quite complex in the study area, since most of the people are engaged in more than one type of work depending on resource availability, work availability and seasonality.

The study identified five major occupational categories in which people of the localities mainly depend for their livelihood. These include day labourer, fisherman, forestry dependent, business and service. Figure 5.1 shows that half of the household heads (50 per cent) mainly worked as day labourers. They had neither land to do agricultural practices upon nor education or money to undertake service or business. As day labourers, they do a large variety of work. They mostly work in the shrimp *ghers*, as it is the main agricultural practice in the study area. Other labour works include paddy plantations, harvesting and soil digging for road construction. In addition, some household heads (mostly stronger persons) went to another part of the country to work as labourers, mainly in the brickfields. Usually

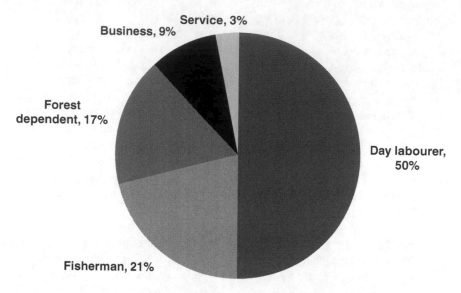

Figure 5.1 Distribution of household heads of the women by occupation

they go this work under the condition to work continuously for six months in the brickfields. However, work as a day labourer is not always available. When they cannot get work to do in the near vicinity of their home, they are usually forced to resort to going to the Sundarbans to catch fish, collect honey, wax, golpata, wood and other available resources.

Approximately 21 per cent of the household heads were mainly involved in fishing. They were also directly and indirectly dependent on the Sundarbans, because the rivers in the Sundarbans are the main source of various species of fish, crab and shrimp fry/fingerlings. The households that are economically stronger have their own shrimp *ghers*, fishing boats, nets and other fishing materials. They also are able to recruit some workers to work in their shrimp *ghers* or in fishing boats to go into the deep forest or deep sea areas during the catching season. The study also found that 17 per cent of the households were mostly dependent on forestry resources and considered this as their main occupation. Going to the Sundarbans, collecting resources (fish, crab, shrimp fry, honey, wax, wood, *golpata*) and selling these to traders or in local markets was considered a vital occupation.

Another 9 per cent of the household heads were involved in various types of businesses. Buying and selling of fish, crab and shrimp fry were the popular businesses in that area. In addition, some of them were involved in commercial motorcycle riding, selling vegetables and fruits that were collected from other parts of the country and small-scale businesses (e.g. grocery shop) in the locality, etc. Only a few (3 per cent) of the household heads were involved in service, mostly in various NGOs, government organisations and teaching in various institutes.

5.1.6 Daily workload of the participants

Similar to many other traditional societies, women in Bangladesh are involved in the domestic household, as well as farming activities, in order to contribute to their livelihood. However, climate-induced environmental and socio-economic changes have increased their burdens and the hardship levels of their work (Islam, 2009). Moreover, household workloads can increase to such a degree that girls can be forced to leave the school to help with domestic chores during and after disasters (Rahman, 2013).

Overall, women work very long days. From waking up in the morning to going to sleep at night, women are involved in a wide range of activities. Women have to perform their home duties, such as: the preparation of food; fetching water; cleaning of the house, clothes and kitchen equipment; taking care of elderly persons and children; chopping firewood; and waste disposal. The study found that 100 per cent of women were engaged in household works. Women do these jobs as a part of the social customs without expecting any economic returns for this labour. The study also found that before Aila in 2009, women spent eight hours per day on average on household work, whereas this increased to nine hours post Aila in 2013 (Table 5.1). This is mainly because women and girls are responsible for collecting and carrying water − a time-consuming and physically demanding task in places where pure drinking water is not easily accessible, such as in the study area.

Women who were engaged in income generation employment worked about 12–16 hours a day on average, including completing their household work. A working woman, Nasima Akter from the Kholpetua village of Gabura union, stated that:

> The typical household work we do at house cannot be done by men. Although I work outside, I need to do all my household work by myself. Therefore, I do not find any free time for myself.

In the study area, 10 per cent of women were engaged in agricultural work before Aila, but this reduced to only 3 per cent after Aila. The workload (average number

Table 5.1 Distribution of working hours of women among major activities

Activity	Percentage of respondents		Average no. of hours per day	
	Before Aila (2009)	*After Aila (2013)*	*Before Aila (2009)*	*After Aila (2013)*
Non-income-generating activities/household works	100	100	8	9
Income-generating activities				
Agricultural works	10	3	4	3
Non-agricultural works	27	23	6	5

of hours) per day in agriculture was about four hours before Aila and it became three hours after Aila. It should be mentioned here that working hours and involvement of women in agriculture had decreased largely due to a decrease in agricultural activities in the study area as a result of various climate change effects. The involvement of women in non-agricultural work was about 27 per cent before Aila, but it decreased to 23 per cent after Aila, because of an overall decrease of income-generating activities in the study area. The women who were engaged in non-agricultural work spent an average of six hours per day before Aila and currently they spend about five hours (Table 5.1). The involvement of women both in agricultural and non-agricultural works indicates that income-generating opportunities have decreased for women in the study area. However, in the meantime, their household workload has increased. Daily workload and its distribution are also closely related to the health status of women. Increased workload indicates more pressure on women's health and less leisure time, which consequently affects their physical and mental health.

5.2 Natural capital

Natural capital is the resource from which people derive all or part of their livelihoods. Natural capital includes land, forests, marine/wild resources, water and air (DFID, 2000). There are various resources that make up natural capital, ranging from intangible public goods, like the atmosphere, to divisible assets used directly in production, like land and trees. Natural capital acquires special importance for those communities where livelihood is heavily dependent on a natural resource base. In the context of the coastal areas of Bangladesh, a particularly close relationship exists between natural capital and the vulnerability context, and this is expected as many of the devastating shocks (such as cyclones, floods) affect natural capital. Consequently, three major natural capitals of livelihood including land, water sources and forest resources are described in detail below.

5.2.1 Land

About 29 million households, which includes about 88 per cent of all households in Bangladesh – live in rural areas. Therefore, for most Bangladeshi people, land and agriculture-based livelihoods are fundamental. Ownership of land determines the status of an individual in rural society (FAO, 2010). This is the only natural capital on which people can form ownership rights. Table 5.2 presents the average size of land holdings and their ownership pattern for the sample respondents. Among the respondents, 97 per cent of the women's households owned a homestead area. The average homestead area was about 0.08 hectare before Aila and reduced slightly (0.07 ha) after Aila. In terms of the ownership pattern, only 3 per cent of the sample women had ownership of their homestead area. For the rest of the households, men held the ownership right for the homestead area. The result also showed that 3 per cent of the sample households did not have any homestead

Table 5.2 Distribution and ownership of land of the respondents' households

Type of land	Land area (in hectares)		Ownership pattern	
	Before Aila (2009)	After Aila (2013)	Male	Female
Average homestead area	0.08	0.07	97%	3%
Average cultivable land*	0.226	0.218	100%	0%
Landless	3%	3%	0%	0%

* The calculation is based on 45 per cent of households who have their own cultivable land.

area of their own. They are considered as landless and usually live by building their dwellings in *khas* land.[3]

In terms of cultivable land, it was found that 55 per cent of the women's households had no cultivable land for crop cultivation. The respondents revealed that due to several devastating climatic shocks, they were forced to sell their land to cope with the adverse situation. Some families sold their land to shrimp *gher* owners due to salinity problems in the surface soil which prevented crop cultivation. The rest of the households (45 per cent) owned some cultivable land; the average land size per household was 0.226 hectares before Aila and was reduced to 0.218 hectares after Aila. In the study area, the cultivable area per household was smaller when compared to the national average of 0.51 hectares (BBS, 2013a). The FGD participants stated that most of the people in this area converted their land to shrimp *ghers*, since other crop production became infeasible because of the salinity. They would have liked to produce crops on their land, but the soil condition did not permit this.

The ownership right of cultivable land solely (100 per cent) belonged to the men of the sample households. Rural women's lack of land rights also limits their access to the other livelihood assets that flow on from the control of land. This statement is not only true for this area, but also almost everywhere in Bangladesh. The government of Bangladesh does not disaggregate any statistics regarding land ownership by gender. Therefore, it is difficult to know exactly how much gender inequality exists in the property distribution of Bangladesh. However, a recent report by Martijin De, Dorothea, and Nicholas (2011) states that women in Bangladesh rarely have equal property rights and rarely hold titles to land. Social and customary practices effectively exclude women from direct access to land.

5.2.2 Water sources

In many developing countries, cultural traditions also make women responsible for collecting water, even when this involves long hours performing heavy physical labour or travelling long distances (Goh, 2012). Therefore, water sources are important in explaining the livelihoods of women, particularly where scarcity of water is a big problem. In the study area, this emerged as a big problem as a consequence of climate change effects in the area. Owing to the salinity problem in the water sources, people used pond water for drinking and other household work before Aila hit the locality in 2009. After Aila, however, the area was inundated with saline water for a long period of time and all sources of freshwater became saline. The situation was such that water was everywhere, but no water could be used due to the increased presence of salinity in the main water sources. A study after Aila found that 86.5 per cent of population had no drinking water sources just after the cyclone (Mallick, 2012). Later, some government organisations and NGOs set up some new hand tube-wells and deep tube-wells and rainwater preservers for the locality. But overall, these were not sufficient to support the high population numbers.

The study found that women were the main collectors of drinking and everyday usage water, and 92 per cent of women said that they were the main collectors of water for their households. For the rest of the women, their husbands usually collected water for households. The average time to collect water varied across households, for some women it was zero as they had access to a nearby source, whereas for some women, it was up to five hours as the nearest water source was quite a distance from their homesteads.

Table 5.3 shows that after Aila, the use of tube-wells, deep tube-wells[4] and rainwater increased, whereas the use of pond water decreased for drinking purposes. Before Aila, 62 per cent of households collected their drinking water from tube-wells, though 41 per cent households were still dependent on ponds for their drinking water. Some households (13 per cent) used rainwater for drinking and quite a few households (1 per cent) used a deep tube-well for their supply of drinking water. The scenario changed after *Aila* and 86 per cent of the households started collecting drinking water from tube-wells, though all of them were not capable of owning a tube-well. Therefore, they had to spend extra hours collecting the drinking water from neighbours' houses. Another 5 per cent of houses started using deep tube-wells for their drinking water.

In addition, 16 per cent of households used rainwater for drinking purposes. For this they needed to store rainwater during the rainy season (about two to three months) in big pots. However, the capacity of rainwater storage is not enough to serve their needs all year round. A household can use rainwater for maybe half of

Table 5.3 Sources of water for the respondents' households

Source of water	Percentage of respondents		Collector of water	
	Before Aila (2009)	*After Aila (2013)*	*Female*	*Male*
Drinking water			92%	8%
Tube-well	62	86		
Deep tube-well	1	5		
Pond	41	10		
River	–	–		
Rain	13	16		
Household water				
Tube-well	14	19		
Pond	84	81		
Deep tube-well	1	1		
River	3	3		

a year and for the other half of the year they need to collect drinking water from other sources. In the dry season when the surface water level goes down and it is not possible to get water from tube-wells, this is an especially miserable time in terms of drinking water supply. In this respect, key informant-1 said that:

> We have made some rainwater reservoirs in the locality. But these are not sufficient to meet the water demand for people. Moreover, the water reserve capacity is limited and can be used only for few months. Eventually, people need to find an alternative source to collect water, most of the time which is located far from the locality and it is women who are responsible to do this hard job.

Table 5.3 also indicates that, for household work purposes, women mainly use pond water. Besides this, they also use tube-wells, deep tube-wells and river water to some extent. The figures are more or less the same as before and after the Aila period. The respondents stated that they did not use any rainwater for their daily household purposes as the rainwater is very precious and only used for drinking purposes.

5.2.3 Forestry resources

The Sundarbans is the source of livelihoods for millions living in its vicinity. Most of them are very poor – have no capital, no land and no water bodies for any type of cultivation. Their livelihood depends, particularly in certain seasons, on shrimp fry collection, crab collection, mussel, clam and oyster gathering, honey and wax collection, and the collection of fuel woods and leaves, all of which is undertaking within the Sundarbans (Biswas, Choudhury, Nishat, & Rahman, 2007; Rahman, 2013).

The study found that 17 per cent of women's households were totally dependent on this forest resource for their livelihood. Other households were also partially dependent on forestry resources for their household income. The main forest-based activities included: fishing, crab collection, wood collection, honey and wax collection and *golpata* collection.

People go to the forest by boat and collect these resources both legally and illegally. This practice is associated with various risks. They have to deal with the danger of attacks by wild animals, such as Royal Bengal tigers, crocodiles and snakes, etc. They also face the possibility of mugging by forest robbers. During the survey, the researcher met some women whose family members had been attacked by both tigers and/or robbers while they went to the forest to collect resources. Still, collecting forestry resources for their livelihoods is the easiest option for the poorest households of the study area, since they do not have to invest any money in these resources. The FGD participants also expressed similar views on this issue. A FGD participant, Alimuddin, stated that:

> Our ancestors used to go to the forest. We follow them. Especially, when there is a shortage of work in the locality, we go to the forest to collect wood, golpata, fish, or crab. Despite the fear of robbery nowadays, we have to go to the forest for survival.

The collection of fuelwood for cooking is another important concern for rural women. Women in the study area absolutely depend on forestry resources for their fuelwood, with 100 per cent of respondents supporting this claim. There are no gas or electricity facilities in the study area. Therefore, a stove made of mud is the only option for cooking and needs fuel like wood, straw or cow dung. However, it has become a great challenge for the women to get wood, as trees have drastically been reduced in the area due to several climate change effects. Straw and cow dung have also become scarce as the pastureland and livestock have become rare in the study area due to the conversion of most of the land areas to shrimp *ghers*. Only 6 per cent and 1 per cent of women occasionally used straw and cow dung respectively for cooking purposes. In general, women were the main users of fuelwood, but it was the male members of the family who were responsible for collecting this from the forest. However, women were responsible for processing the wood by chopping where necessary, for cooking purposes.

5.3 Financial capital

The level, variability and diversity of income sources and access to other financial resources (e.g. credit, savings) together contribute to the formulation of financial capital. Financial capital is especially important for the livelihood groups who live in a climate volatile area, as it can be used as collateral against risks and shocks (Ellis, 2000). The study describes the financial capital of the respondents in terms of income and expenditure structure, savings and credit availability.

5.3.1 Income structure

Most of the households in the study area were engaged in more than one income-generating activity that together produced their total annual income. The main income sources for women's households included shrimp or fish farming, labour wages, forestry resources, business, agriculture and service. Most of the women's households (67 per cent) were engaged in daily labour work. However, they could not find work on a regular basis and so the wage rate was very low. The average annual income from wage labour was estimated at approximately US$ 545 before Aila and US$ 571 after the Aila period. The wage income increased slightly from 2009 to 2013, but this was mainly because of ongoing inflation, rather than increased work availability. Therefore, in practice, it did not help to increase the purchasing power of households.

Shrimp farming or fish farming was considered to be the most lucrative sector for income generation. The average annual income of households from shrimp/ fish farming was calculated at US$ 1077 in 2009 and US$ 780 in 2013. The participants in the FGDs mentioned that due to a breakdown of embankments, in general, shrimp *ghers* were washed away and an attack of various infectious diseases to the shrimp industry in the Aila year caused huge losses. Though the income has reduced from this industry more recently, more households (67 per cent) were involved in this sector in 2013 than in 2009 (65 per cent) since the land became saline, and there is now nothing productive that can be done with this land except shrimp farming.

Another major income-generating sector for the households was income from forests. In almost every household, one or more members went to the forest to collect forestry resources and to sell these to the local markets as part of their living needs and overall livelihoods. Unfortunately, the average annual income from this sector has also reduced since Aila (US$ 468 in 2009 to US$ 388 in 2013). The involvement of households also slightly reduced (29 per cent to 27 per cent) during this time period. The main reasons for this decline reported by the FGD participants were the increased dangers of robbery in the Sundarbans and the decreasing amount of resources available in the forest. The involvement of the households in business (e.g. trader of fish and agricultural products, grocer, tailor, wood dealer, shopkeeper) was the same (31 per cent) both before and after Aila periods. However, the average annual income from this sector decreased to US$ 662 in 2013 from US$ 740 in 2009.

Although agriculture is the main occupation for rural Bangladeshis, this is a problematic income-generating sector in the study area due to various climatic factors. This sector includes income from crop production, homestead gardening, fruits and vegetables gardening, livestock and poultry rearing, etc. As mentioned earlier though, due to various climatic reasons, this sector is the lowest income-generating sector of the area. The contribution of this sector in households' income is very low both in terms of value adding and involvement. The average annual income from this sector was only US$ 213 in 2009 and US$ 140 in 2013. The percentage of households involved in agricultural production was only 29 per cent in 2009 and this reduced to only 19 per cent in 2013. The main reason for the reduction in the agricultural activities in the study area was the presence of salinity in the surface soil. Unavailability of fodder crops, pasturelands and poultry feed were other reasons for reduction in poultry and livestock production.

The service sector (e.g. service in any company outside their village, in NGOs, etc.) involved the least number of households in the study area. Only 7 per cent of the respondents' households were involved in service in 2009, but that increased slightly in 2013 to 9 per cent. The average annual contribution of this sector to the households' income was estimated at US$ 570 in 2009 and US$ 572 in 2013. Figure 5.2 present the income information for the respondent households. The total average annual income per household was estimated at US$ 1657 before Aila in 2009 and US$ 1383 after Aila in 2013, which was low when compared to the national average (US$ 1776) (BBS, 2011). The figures also indicate that households were, comparatively, in a disadvantaged position and became poorer due to climate change shocks.

5.3.2 Expenditure structure

The present study also collected data on women's household expenditure in two categories, food items and non-food items. The annual average expenditure on food items was estimated at US$ 718 in 2009 and US$ 738 in 2013. Conversely,

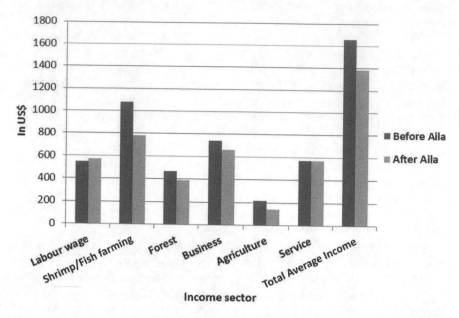

Figure 5.2 Income distribution of women's households by sectors

the annual average expenditure on non-food items was estimated at US$ 290 in 2009 and US$ 334 in 2013. As a result, the total annual expenditure per household increased slightly by 2013 (US$ 1072) compared to before Aila occurred in 2009 (US$ 1007) (Table 5.4). Both the figures are much lower than the national annual expenditure per household (US$ 1711) (BBS, 2011), which indicates a lower purchasing power of households in the study area.

The results also indicate that the households spent most of their income on food items, which is an indication of the poor livelihood of the respondents. The share of food items and non-food items in total household expenditure were calculated, about 70 per cent and 30 per cent respectively on average, both before Aila and after the Aila period.

5.3.3 Balance of income and expenditure

The study identified the respondents' financial status by balancing their income and expenditure. The results reveal that about 37 per cent of the respondent households experienced 'income surplus' in 2009, but that figure decreased to 17 per cent in 2013. On the other hand, a break-even/no saving situation was the most common situation for the households and this increased to 70 per cent in 2013 from 59 per cent in 2009. Income less than expenditure or 'income deficit' increased to 13 per cent in 2013, after sitting on 4 per cent before *Aila*. These estimates give a clear indication of the overall poor financial status of households and

Table 5.4 Household expenditure structure

Expenditure	Average annual expenditure (in US$)	
	Before Aila (2009)	*After Aila (2013)*
Food items	718	738
Non-food items	290	334
Total	1007	1072

a worsening of the situation after Aila. The households largely lost their assets and income opportunities and, as a result, they became marginalised. The estimates also indicates that women's households could not cope well with the devastating damage of Aila and other climatic consequences, as they could not save income to be used as collateral at the time of disasters. The situation remained the same in 2013. A respondent named Horidasi from the Sonakhali village of Padmapukur union said:

> We cannot think of saving money. With the little income of my husband we cannot even cover our basic expenses. Every month there is some deficit and I need to borrow money to run my family.

5.3.4 Credit availability to the participants

Credit is considered to be an important financial asset for rural households, considering its impact on socio-economic development. In Bangladesh, microcredit is a very popular concept and it is mainly given to rural women to improve their livelihoods. At the time of data collection, it was found that in the study area, 32 per cent of the respondents had not taken any kind of credit from any source during the last five years. The reason behind this is some of them did not like to take credit from others, although they were in need, and some of them had not found any suitable source of funding. Alternatively, 68 per cent of the respondents received credit from various types of sources to meet their needs. The proportion of women who took credit was much higher than the national average for rural areas, which is 35.08 per cent (BBS, 2012). The amount of credit received by the respondents ranged from US$ 52 to US$ 3847. Among the respondents, only 14 per cent of women received credit more than US$ 642. The distribution of respondents according to the amount of credit received is presented in Table 5.5. Besides their needs, the respondents also had to consider their repayment capacity when taking credit.

The study also investigated the reasons why the respondents took credit. The results revealed that most of the respondents (44 per cent) took credit for family purposes. These included: food expenditures, children's education, the making and repairing of the house, as a dowry for a daughter's marriage, monies paid to other credit organisations and buying some poultry and livestock, etc. On the other hand, 26 per cent of the women took credit for business purposes. They

Table 5.5 Amount, reasons and sources of credit for respondents

Amount of credit (in BDT)	Percentage of respondents	Reason for credit	Percentage of respondents	Source of credit	Percentage of respondents
No credit	32	Family purposes	44	NGO	54
Up to 5,000	11	Business	26	Bank	15
Up to 10,000	15	Fishing	16	Relatives	19
Up to 20,000	11	Medical	10	Co-operative society	9
Up to 50,000	19	Agriculture	4	Others	3
> 50,000	14	Others	–	–	–

usually gave this money to their husbands or sons to start some small businesses, such as opening a grocery shop or investing in a small trading business. Another 16 per cent of the women used their credit to cultivate fish and to buy some fishing equipment. Furthermore, 10 per cent of the respondents took credit for medical purposes and only 4 per cent of the respondents used their credit for the improvement of agricultural production. A key informant-9 said in her interview that:

> From my experience I have seen that women mostly take credit to bear their family expenses. Now, we are trying to motivate them to invest their credit in small businesses. We also give women training on making handicrafts so that they can use this credit in proper way and pay the installments without hassle.

The study also identified the sources of credit for the women in the study area. The main source of their credit was various NGOs. The study area was identified as one of the most vulnerable regions to climate change worldwide. Therefore, many NGOs have started working in this region. The area came more into focus after Cyclone Aila, which had a disastrous impact on that area. Afterwards, the NGO activities became more prominent in supporting the livelihoods of the people in that area. Many international organisations then also provided funding through these NGOs.

The NGOs who provided credit to the women in the study area included the Nowabenki Gonomukhi Foundation (NGF), Shushilon (a local NGO), BRAC (an international development organisation based in Bangladesh), Caritas Bangladesh and Borsha (a local NGO). Most of the respondents (54 per cent) received credit from local, national and international NGOs. Others sources of credit included relatives (19 per cent), banks (15 per cent), co-operative societies (9 per cent) and others (3 per cent). The banks who gave credit to that area included the Grameen Bank,[5] the Bangladesh Krishi Bank[6] and the Al-Baraka Bank.[7] Although the amount of credit was not high, most women felt happy that at least they had some sources from which they could receive financial support during times of struggle.

5.4 Social capital

Although there is much debate about what is exactly meant by the term 'social capital' and the aspects it comprises, quite often access and amount of social capital is determined through family and gender relations, social relationships and networks and social groupings. Social capital is an important factor in understanding a community's livelihood, because it helps understand the relationship between assets, institutions and livelihoods (Bebbington, 1999). This capital can positively or negatively impact on the livelihoods of women depending upon its composition. Social capital is explained in this study according to two categories: socio-demographic profile and social networking, which are described below.

5.4.1 Socio-demographic profile

The household structures of respondents are discussed in this category under the following sub-headings.

5.4.1.1 Family size and gender composition

The household size of the respondents was found to be high (5.05 persons per family) compared to the average household size of 4.6 persons per family in Bangladesh (NIPORT, 2015). The average number of female members (2.44) in the family was slightly higher than that of male members (2.34). The households were mostly male-headed (92 per cent). Only 8 per cent of households were headed by women, who were mostly widowed. The national average for male- and female-headed households is 89 per cent and 11 per cent respectively (NIPORT, 2015). These numbers are representative of the male-dominant social structure of Bangladesh, which is strongly evident in the study area, with low contributions of women in household decision-making processes.

5.4.1.2 Dependency ratio

The dependency ratio indicates the proportion of family members who are typically not in the labour force. It is used to measure the pressure on a productive population. The dependency ratio for the respondents' households was calculated at 0.664, which indicated more than half of the family members were dependent. Again among the dependent members, the majority were female. This result indicates that women have less productivity in terms of generating income. This is not because that they are unable to do productive work; it is the socio-cultural restriction and their family responsibility that restricts them from engaging in such work.

5.4.1.3 Religion

The results indicated that most of the respondents (88 per cent) belonged to the religion Islam. The rest of the respondents (12 per cent) belonged to the Hindu religion. The national distribution of population in Bangladesh is: Islam (89.7 per cent), Hinduism (9.2 per cent), Buddhists (0.7 per cent) and Christians (0.3 per cent) (BBS, 2013a). The people in the rural area were very much devoted to their respective religion, which increased the patriarchy in the society.

5.4.2 Social networking

Social networking helps in building social connections that enable and encourage people to mitigate the effects of shocks or lack of other capitals through informal networks and developing social cooperation. The two aspects, namely migration and networking with local government organisations, are described under this category.

5.4.2.1 Migration to other places

Migration is a social behaviour that can occur for various reasons; however, in the study area it related to the consequences of climate change. The study found that only 27 per cent of the respondents' family members migrated to other areas of the country or overseas. They mostly migrated to earn money or to find jobs as there was scarcity of work in the vicinity after several natural calamities. Among the migrated people, the highest proportion (15 per cent) of people migrated to other districts (inter districts) of the country. Most of them migrated to work in the brick-fields. Besides this, 5 per cent of them migrated to nearby inner upazilas of the Sat-khira district. About 4 per cent of the family members of the respondents migrated to the capital city Dhaka for a job, and only 3 per cent migrated overseas for a better life or as manpower. The FGD participants claimed that there was a shortage of work in the local vicinity. A FGD participant Aklima Bagum stated that:

> We do not have much work opportunities in the locality, especially after Aila. Therefore, physically strong men go to work in other places. At that time, we (women) have to take care of our family and bear more responsibilities to maintain our family.

The study also found that 73 per cent of the respondents' family members had not migrated to other places. Some of them did not get enough opportunity to migrate to other places and some of them were not interested in migrating to other areas from their place of origin. The study also found that about 3 per cent of the respondents' relatives or family members had permanently migrated to other areas of the country, as they lost all of their livelihood opportunities due to various climate change effects (Table 5.6).

5.4.2.2 Networking with local government organisations

The local government organisations that work in the study area are Union Pari-shad and Upazila Parishad, which include all government organisations at the upazila level. Rural people usually seek assistance from the Union Parishad

Table 5.6 Migration status of respondents' family members

Destination to migrate	Migration status		Permanently migrated
	Yes (27%)	No (73%)	
Dhaka (capital city)	4%	Most of the people did	
Overseas	3%	not migrate to other	3%
Inter districts	15%	places due to lack of	
Locally	5%	opportunities.	

if they face any problems. The responsibilities of the Union Parishad include: assisting and cooperating in the development of education and health; the construction and maintenance of local roads, water sources and embankments; the peaceful resolution of local disputes; the promotion of social resistance over violence against women, terrorism and all types of crimes; and an increase in awareness of the needs of women and children, taking concrete actions where necessary (GoB, 2015).

In the study area, men usually take the responsibility to make contact with local government offices. Women may be interested but due to the social and culture norms, there is no opportunity to engage with such contacts. Therefore, women's access or networking with this organisation was limited. Only 25 per cent of the women had adequate access to these organisations. About 35 per cent of the women had limited access to these organisations, whereas 40 per cent of the women had no access to these organisations. The situation was worse before Aila (2009). At that time 20, 30 and 50 per cent of women had adequate, limited and no access to the local government organisations respectively. Accessibility of women to different welfare facilities that are also related to social networking is further discussed in Chapter 6.

5.5 Physical capital

Physical capital comprises the basic infrastructure and producer goods needed to support livelihoods. Physical capital represents a stock of plants, equipment, infrastructure, sanitation system and other productive resources that are owned by individuals or households (Bebbington, 1999). It plays an important role in accelerating the livelihood of households in a positive direction. Physical capital such as housing conditions, productive assets (livestock, poultry, fishing boat and nets, trees, sanitation facilities) are examined in this section.

5.5.1 Housing conditions

Shelter is a basic need for livelihoods and can be represented largely by the condition of the house in households. The type of house indicates the wellbeing or status of the household. Therefore, the present study investigated the housing conditions of the respondents, and compared this before Aila and after Aila. Houses were categorised into four types such as: *pucca* house, semi-*pucca* house, iron sheet house and *kacha* house. *Pucca* houses are made out of brick, cement, iron bars etc., while semi-*pucca* houses have walls that are made from the same materials, but the roofs are made from iron sheets. The main structure of iron sheet houses is made from wood or bamboo that is wrapped with iron sheets. *Kacha* houses are mainly made from *golpata* of the Sundarbans, with the help of bamboo or wood. Some of the houses are also made from straw and mud.

In rural Bangladesh, one household may have one type of house or different types of houses. For example, one household may build one or two rooms as a *pucca* or semi-*pucca* house, another one or two rooms as an iron sheet house, and the kitchen as *kacha* house. In the study area, most of the households had one house structure and a separate kitchen. The results showed that 97 per cent of the respondents' houses had been fully or partially destroyed at the time of the devastating Aila. Strong winds followed by a high tidal surge were the main destroying reasons for this. After Aila, water remained for a long time in the houses and all types of houses were severely affected by saline water. Before Aila, about 78 per cent of the respondents owned at least one *kacha* house. After Aila, 75 per cent of the respondents had at least one *kacha* house and most of the *kacha* houses had been remade after being previously destroyed during Aila.

During the survey period, 71 per cent of the respondents had at least one iron sheet house, whereas only about 51 per cent of the respondents had a one iron sheet house before Aila. Only 11 per cent of the respondents had one semi *pucca* house in 2013, whereas 8 per cent of the respondents had one semi *pucca* house in 2009 before Aila. Very few households (5 per cent) had one *pucca* house in the data collection time, whereas only 3 per cent of the respondents had one *pucca* house before Aila. In rural Bangladesh, usually people who are considered 'rich' own a *pucca* house and people who are poor own *kacha* houses. Thus, the structure of the house demonstrates the economic status of a household, and for the present study, the housing condition demonstrated the poor livelihoods of the respondents.

The study found that all categories of houses had increased in terms of quantity after Aila because of the various rehabilitation activities of many international, national and local NGOs. In terms of the ownership pattern, only 2 per cent of houses were owned by women in the study area, whereas 98 per cent of the houses were owned by the men, which indicates that women are largely deprived in terms of physical asset endowments.

5.5.2 Productive assets structure

Physical assets in the form of productive assets are said to be associated with the increased income and better livelihoods of households. Productive assets of a rural household include: poultry, livestock, agricultural equipment, fishing nets and boats, trees, homestead vegetables gardens, radios, televisions, solar power, refrigerators, jewellery, etc. In the study area, it was found that most of the people were poor and as such, agricultural activities were few due to soil salinity problems. Thus, the most common physical assets of the respondents' households were limited to livestock, poultry, fishing boats and nets and trees, which could explain their present livelihoods in the study. About 38 per cent of the respondents had livestock in their family before Aila and the average number of livestock was six per household, whereas after Aila it came down to about 30 per cent of respondents having an average number of livestock of two per household. Only 4 per cent of the women had ownership of livestock during the time of data collection (Table 5.7).

Poultry is the most popular asset for women in rural households. Before Aila, about 87 per cent of the respondents had poultry, the average being 12 per household. However after Aila, both the percentage of households and average number of poultry per household had decreased. After Aila, about 80 per cent of the households only owned an average of five poultry per household. The average number of poultry per household had decreased significantly after Aila. During Aila, most of the poultry had washed away with the tidal surge and then after Aila, due to a lack of poultry feed and space in the homestead area, households found it difficult to raise poultry. The respondents mentioned that after Aila, they tried to buy and raise poultry and livestock with the help of some local NGOs, but livestock and poultry feed was a problem for them as there was no pastureland or fodder crops grown in that area. In terms of ownership of poultry, 36 per cent of the respondents owned this asset, which was the highest percentage for women in any ownership pattern within the households. The high numbers of participation and ownership of poultry indicates that women are very comfortable in raising poultry and want to contribute to this sector for the betterment of their family.

Fishing nets and boats are also important household assets for the coastal community, since most of the households engage in fishing in different forms. Among the respondents' households, 43 per cent of the household had owned a fishing boat and net before Aila, but this came down to 33 per cent of households in 2013. The average number of fishing nets and boats were the same (two) during both time periods. Only 2 per cent of fishing materials were owned by women (Table 5.7).

A large number of households (65 per cent) had trees before Aila and on average the number of trees was about 12 per household. However in 2013, the number of households (19 per cent) and the average number of trees per household (only three) had been reduced by soil salinity. It was observed during the time of the data collection that most of the households had no trees at all in their homestead. Some of them mentioned that they had tried to plant new trees but that they failed to grow. Very few of them had succeeded in plantation growth. No homestead gardens were found during the time of data collection but, most FGD participants mentioned that a couple of years ago their homesteads had held various types of vegetables, fruits, other trees, livestock and poultry, but Cyclone Aila had destroyed most of their assets and they failed to regain these, mainly because of ongoing heavy salinity and water logging problems.

Among the other household assets, 26 per cent of the respondents' households owned a bicycle or, a rickshaw van or a motorcycle. Only 9 per cent of

Table 5.7 Distribution and ownership pattern of major physical assets

Types of asset	Percentage of households		Ownership pattern in 2013 (percentage)	
	Before Aila (2009)	After Aila (2013)	Male	Female
Livestock	38	30	96	4
Poultry	87	80	64	36
Fishing boat and net	43	33	98	2
Trees	65	19	97	3

the respondents' households owned a solar panel. Solar panels were their only source of power supply as there was no electricity supply in the study area. People obtained these solar panels from different NGOs at discounted prices. Besides this, 32 per cent of respondents' households owned either a radio or a small television. Among the respondents, only 2 per cent had gold or silver jewellery. There was no household found that owned any agricultural equipment or refrigerator. Table 5.7 presents the distribution and ownership of major household assets.

5.5.3 Condition of sanitation facilities

Sanitation is a fundamental human right that safeguards health and human dignity and, hence, is a key indicator of livelihood. Lack of sanitation is a public health disaster and more embarrassing for women. The present study investigated the types of toilets used by the households in the study area to get an idea about their livelihood wellbeing. It is notable that every household now has a toilet in the study area. That happened because of a campaign by the government and NGOs to help rural communities. After Aila, several NGOs made toilets in many households in the study area. Most of the households used semi *pucca* toilets (75 per cent), whereas some used *pucca* (16 per cent) and *kacha* (9 per cent) toilets during data collection. *Pucca* toilets are made out of bricks, semi-*pucca* toilets are made out of wooden and iron sheets, and *kacha* toilets are made out of bamboo and *golpata*.

5.6 Conclusion

This chapter has presented the status of five livelihood capitals of women and their families in the study area. These capital endowments were the indicators of women's livelihoods which were determined by several livelihood strategies performed by them and resulted in positive or negative livelihood outcomes. The distribution of capitals was also heavily influenced by the context of vulnerability, which was described from the point of climate change effects such as Cyclone Aila. Therefore the above discussion gives an indication of asset compositions, their changes and other issues related to livelihood scenarios. The next chapter will discuss the key areas of vulnerability in relation to climate change. It will identify the impacts of climate-related issues on women's livelihood and out-line women's responses to and coping strategies against these vulnerabilities. It will also consider what accessibility to different facilities might mean in helping reduce vulnerabilities overall.

Notes

1 Practice of collecting shrimp fingerlings from wild sources (e.g. rivers, canals).
2 A homestead is land owned or occupied by a dwelling unit of a household, along with the adjoining area.
3 *Khas* land is state-owned land and legally reserved for distribution to landless people.
4 From an engineering point of view, when a tube-well penetrates at least one imperme-able layer, it is known as a deep tube-well. But in Bangladesh, when a tube-well is deeper than 75 metres, it is called a deep tube-well as per the Department of Public Health Engineering (DPHE).
5 The Grameen Bank is a Nobel Peace Prize-winning microfinance organisation and com-munity development bank founded in Bangladesh. It makes small loans known as micro-credit to the impoverished without requiring collateral.
6 Bangladesh Krishi Bank (BKB) is a 100% government-owned specialised bank in Bang-ladesh. *Krishi* means agriculture. Since its inception, BKB has provided remarkable lev-els of financing to the agricultural sector. BKB also performs commercial banking.
7 A commercial bank operating under Islamic (Halal) principles.

References

Asaduzzaman, M., Reazuddin, M., & Ahmed, A. U. (Eds.) (1997). *Global climate change: Bangladesh episode*. Dhaka: Department of Environment, Government of the People's Republic of Bangladesh.

BBS (2011). *Bangladesh: Household Income and Expenditure Survey 2010*. Bangladesh Bureau of Statistics, Statistics and Informatics Division, Ministry of Planning, Govern-ment of the People's Republic of Bangladesh, Dhaka.

BBS (2012). Community Report-Satkhira Zila, *Population and Housing Census 2011*. Bangladesh Bureau of Statistics, Statistics and Informatics Division, Ministry of Plan-ning, Government of the People's Republic of Bangladesh, Dhaka.

BBS (2013a). Statistical Yearbook of Bangladesh-2012. Bangladesh Bureau of Statistics, Statistics and Informatics Division (SID), Ministry of Planning, Government of the Peo-ple's Republic of Bangladesh, Dhaka.

BBS. (2013b). *Literacy assessment survey (LAS)-2011*. Dhaka: Bangladesh Bureau of Sta-tistics, Statistics and Informatics Division (SID), Ministry of Planning, Government of the People's Republic of Bangladesh.

BBS. (2013c). *Gender statistics of Bangladesh 2012*. Dhaka: Bangladesh Bureau of Statis-tics, Statistics and Informatics Division (SID), Ministry of Planning, Government of the People's Republic of Bangladesh.

Bebbington, A. (1999). Capitals and capabilities: A framework for analyzing peasant via-bility, rural livelihoods and poverty. *World Development*, 27(12), 2021–2044.

Biswas, S. R., Choudhury, J. K., Nishat, A., & Rahman, M. (2007). Do invasive plants threaten the sundarbans Mangrove forest of Bangladesh? *Forest Ecology and Manage-ment*, 245(1–3), 1–9.

CPD and MJF. (2014). *Estimating women's contribution to the economy: The case of Bang-ladesh*. Presented at the dialogue on 'how much women contribute to the Bangladesh economy', results from an empirical study organised by Centre for Policy Dialogue (CPD) in partnership with Manusher Jonno Foundation (MJF), May 05, 2015, Dhaka.

DFID. (2000). *Eliminating world poverty: Making globalisation work for the poor* (DFID White Paper). London: Department for International Development.

Efroymson, D., Biswas, B., & Ruma, S. (2007). *The economic contribution of women in Bangladesh through their unpaid labour.* Dhaka: WBB Trust and HealthBridge.

Ellis, F. (2000). *Rural livelihoods and diversity in developing countries.* Oxford: Oxford University Press.

FAO. (2010). *On solid ground: Addressing land tenure issues following natural disasters eroding rivers, eroding livelihoods in Bangladesh.* Rome: Food and Agriculture Organization of the United Nations.

GoB (2015). Local Government Division Website, Ministry of LGRD and Co-operatives. Government of the People's Republic of Bangladesh, Dhaka. Retrieved from http://www.lgd.gov.bd/

Goh, A. H. X. (2012). *A literature review of the gender-differentiated impacts of climate change on women's and men's assets and well-being in developing countries* (CAPRi Working Paper No. 106). Washington, DC: CGIAR Systemwide Program on Collective Action and Property Rights (CAPRi).

Islam, R. (2009). *Climate change induced disasters and gender dimensions: perspective Bangladesh.* Peace and Conflict Monitor, University of Peace, Costa Rica.

Kollmair, M., & Gamper, St. (2002, September 9–20). *The sustainable livelihoods approach.* Input paper for the integrated training course of NCCR North-South Aeschiried, Switzerland, Development Study Group, University of Zurich (IP6), Zurich.

Mallick, B. (2012). *The way society deals with increasing vulnerabilty: Socio-economic analysis for vulnerability-oriented spatial planning in coastal zone, Bangladesh* (Unpublished Ph.D dissertation). Submitted to the Karlsruhe Institute of Technology (KIT), Germany.

Martijin De, H., Dorothea, H. M., & Nicholas, P. (2011). *Bangladesh: Food security and land governance factsheet* (7 pp.). Utrecht: The IS Academy on Land Governance for Equitable and Sustainable Development (LANDac).

NIPORT. (2015). *Bangladesh demographic and health survey 2014: Key indicators.* Dhaka: National Institute of Population Research and Training (NIPORT), Ministry of Health and Family Welfare, People's Republic of the Government of Bangladesh.

Rahman, S. (2013). Climate change, disaster and gender vulnerability: A study on two divisions of Bangladesh. *American Journal of Human Ecology, 2*(2), 72–82.

Rakodi, C. (1999). A capital assets framework for analysing household livelihood strategies: Implications for policy. *Development Policy Review, 17*(3), 315–342.

Scoones, I. (1998). *Sustainable rural livelihoods: A framework for analysis* (IDS Working Paper 72). Brighton, UK: Institute of Development Studies.

UNESCO. (1997). *Educating for a sustainable future: A transdisciplinary vision for concerted action.* UNESCO teaching and learning for a sustainable future. Paris: United Nations Educational, Scientific and Cultural Organization (UNESCO).

6 Vulnerability of women's livelihoods and the coping mechanisms to address climate change impacts

The impacts of climate change are likely to enhance the vulnerability of many communities, particularly those that are already susceptible to climate variability, as well as development pressures, such as the rural women of developing countries. The purpose of this chapter is to understand the vulnerability of women's livelihoods to climate change impacts. The concept of the Disaster Crunch Model (DCM) is followed to obtain the above objective. In the context of the present study, the purpose of using this model is to identify how several hazards affected women in the study area differently; the specific aspects of gender relations and livelihood vulnerabilities; and women's specific needs, concerns, priorities and their own coping mechanisms in reducing risks and associated vulnerabilities from disasters. The knowledge gained from these insights will provide a more nuanced understanding of issues as seen from the perspective of rural women; it will also allow for more specific targeting of programmes that assist in addressing gender and climate change issues overall.

Given this focus, the following sections discuss the key vulnerabilities of women's livelihoods due to the impacts of climate change and will present a summation of their coping strategies against these vulnerabilities. The chapter concludes with a discussion of the accessibility of women to various welfare facilities that can potentially reduce their climate-induced vulnerability and enhance their long-term adaptive capacity.

6.1 The progression of women's vulnerability

> Women are vulnerable not because of natural weakness (i.e., because of their sex), but rather because of the socially and culturally constructed roles ascribed to them as women (i.e. because of their gender).
>
> (UNDP, 2010, p. 15)

The present study started assessing the vulnerability of women by asking the question, "Do you think women are more vulnerable than other groups in society?" to all personal interviewees, focus group discussants and key informants. All the respondents agreed that women were more vulnerable than other groups in the society. According to the DCM, women's vulnerability is progressed due to various root causes, dynamic pressures and unsafe conditions. In light of this, this section gives a précis of the factors that were responsible for the progression of women's vulnerability. Accordingly, this section (6.1) briefly discusses the sources of vulnerabilities of women based on overall consultation with key informants and FGD participants of the study and related secondary sources.

In a poor and underprivileged community like the study areas, climate change has a disproportionately greater effect on women. The prevailing socio-cultural environment further increases women's vulnerability to climate change impacts. The major factors that put women into more vulnerable conditions are:

i) **Limited access to resources**: In rural Bangladesh, women have limited access to key resources such as land, livestock, houses, machinery and savings. Access to land and security of tenure is often highlighted as an important cause of women's vulnerability (Agarwal, 2003; Jacobs, 2002). The results of this study also found similar statistics such as women had no ownership of cultivable land (see Chapter 5). Although women work on croplands, they have limited or no control over the land, as they do not own it and therefore cannot make decisions regarding its use. Thus, there is no personal security attached to it. Women's limited access to resources was also revealed for other household assets (see Chapter 5). Men retain greater access to each form of the five livelihood capitals: human, natural, financial, social and physical. This allows men to diversify their livelihoods and adaptive capacities to climate change to a much greater extent than women.

ii) **Dependence on natural resources and related work responsibilities**: Women in rural areas in developing countries are especially vulnerable when they are highly dependent on local natural resources for their livelihoods (UN WomenWatch, 2009). As the primary users and managers of natural resources (being typically responsible for fetching water and fuelwood, and bringing it to the house, for example), women depend on the resources most at risk from climate change (UNDP, 2010). High dependency on agriculture, forestry resources, fisheries and biofuels increases women's vulnerability. Climate change effects revealed in the study area include changes in the availability of natural resources, such as drinking water, wood, forestry products and fisheries, which potentially affect the livelihoods of women.

iii) **Lack of education and access to information**: In Bangladesh, women have typically received fewer school years and grown up as less educated persons. Traditionally, the role of girls in Bangladesh has been linked to the households; early marriage, cultural norms and religious orthodoxy have also been responsible for the low educational attainments of women in Bangladesh (Hossain & Tisdell, 2003). In the national literacy survey, it was found that only 35 per cent of rural women continued their study after primary-level education (BBS, 2013b). For the present study, this figure was even worse – only 27 per cent of women continued their study after primary level (see Chapter 5). Due to their limited education, women are at a disadvantage, as they have less access to crucial information and fewer means to interpret this information. This can affect their ability to understand and act on information concerning climate risks and adaptation measures. Limited education also shrinks their opportunities to become involved in income-generating activities and, therefore, increases their vulnerability.

iv) **Limited mobility:** Women in rural Bangladesh are less mobile due to strict and gendered codes of social behaviour and, as such, have lesser chances to escape from affected areas (CCC, 2009). They are restricted from leaving their houses to seek safety and shelter alone and without the permission of male head of the household. Remaining at home can leave them vulnerable in two ways: firstly, they stay when climate change events such as cyclones hit hard if there is no male member present during the time of disaster; and, secondly, if the shelter centre is far from their house, they also avoid going there without permission and thus face a worst case scenario – a backlash from the community. This is one of the reasons that the death rate of women was higher in all cyclones in Bangladesh, as their powerlessness in life was seen to override their survival instincts. It is also because of this immobility that they miss out on opportunities for alternative livelihoods, as they are prevented from migrating to other places.

v) **Limited roles in decision making:** Women have no role in the decision-making processes of households and community-level governance in rural Bangladesh, which impacts generally on their lives, environment and aspirations (Kabeer, 2005). A women's role in the household is mainly seen as reproductive labour (domestic work, child care and care for the sick and elderly), whereas men's roles are seen as productive labour (paid work, self-employment and subsistence production), and this ultimately creates the differences in the decision-making process by putting men in a position of power.

vi) **Sex disparity:** In Bangladesh, there are gender inequalities with respect to: the enjoyment of human rights; political and economic status; land ownership; housing conditions; exposure to violence; education; and health (in particular reproductive and sexual health). All of these make women more vulnerable before, during and after climate change-induced disasters (Baten & Khan, 2010). Thus, women are disproportionally affected by the adverse effects of climate change in regards to: agriculture and food security; water resources; human health; human settlements and migration patterns; energy; and a lack of transport and communication mostly, because of their gender and lack of identity. As a result, they face much greater challenges in livelihood strategies to adapt to the changing environment than men and most women in developed countries.

6.2 Climate change and women's vulnerability

Climate change can be defined as extreme events or as a continuous process. The respondents of the personal interviews, focus group discussions and key informant interviews of this study were asked about the climate change events or the climatic hazards they had experienced during the past years. Since most of the respondents were local people, they observed the gradual changes of climate in that area. According to them, the common climatic events or hazards that they had observed in their locality were increased intensity of storms and cyclones, increased soil salinity, increased temperature, water logging and floods followed by heavy rainfall. During a FGD in East Patakhali village, a participant Hajera Begum stated that:

> Our environment has changed. Now we feel excessive heat and heavy rainfall. Frequency of cyclones and floods has also increased. About 20 years ago the situation was not the same. Now we are in a greater risk. Salinity is also causing a lot of troubles in our life.

The impacts of climate change events are interlinked and they create vulnerability in women's lives in various interrelated ways. The degree of vulnerability varies from person to person. Therefore, the present study has categorised the degree of vulnerability into four categories: severe, moderate, low and no vulnerability. The major areas of vulnerability of women's livelihood are described below. The percentages of women in relation to their vulnerability to different aspects are also presented in Table 6.1.

6.2.1 Income vulnerability

Women's income is more likely to be derived from the informal sector, which is often the worst hit by disasters and the least able to recover (Rahman, 2013). In the rural areas of Bangladesh more generally, like those of the study area, there exist few opportunities for income generation for people. In general, women represent the majority of low-income earners and those in the most disadvantaged situation regarding income vulnerability. In this study, 62 per cent of the women said that their income sources had become severely vulnerable due to recent climate change effects. Another 10 per cent of the women felt that their income sources were moderately vulnerable. However, 25 per cent of the women also responded that there was no vulnerability in their income sources, the reason

Table 6.1 Degree of vulnerability due to climate change impacts on women's livelihoods

Vulnerability indicator	Severe	Moderate	Low	No vulnerability
Income source	62%	10%	3%	25%
Household assets	89%	10%	1%	0%
Lives and health	50%	39%	9%	2%
Food security	70%	13%	0%	6%
Education	6%	5%	2%	87%
Water sources	34%	3%	20%	43%
Sanitation	13%	5%	4%	78%
Shelter and security during times of disasters	79%	16%	2%	3%
Communication and transportation	52%	38%	3%	7%

being that these particular women were not engaged in any income-generating activities at all (Table 6.1).

A personal interviewee Fatema Begum from Khalishabunia village of Gabura union said that:

> I had 20 chickens and 15 ducks from which I earned some money by selling their eggs. I had also a vegetable garden. But Aila came as a curse to our lives which washed away all my poultry, and destroyed my garden. It is four years since Aila hit us, but still I could not recover from my losses. I had no ability to buy these poultry again. I tried to grow some fruits and vegetables in my homestead, but because of high salinity after Aila, I failed to do so.

Discussion with women through personal interviews and FGDs revealed that, in the study area, although women mostly practised homestead-based livelihoods, the few income generation opportunities they had were lost due to natural disasters such as Cyclone Alia. The deaths of livestock and poultry during disasters increased the vulnerability of women's livelihoods, as these are one of the main income-generating sources for women. During and after disasters, the lack of fodder for livestock and poultry resulted in reduced milk and meat production, which overall worsened the income situation for women. Moreover, cyclones such as Aila also reduced the employment opportunities available for women who had previously worked in agricultural fields. Many of the respondents used to practise homestead gardening from which they could earn some money by selling excess fruits and vegetables. The recent climatic events, including increased soil salinity, however, have constrained them in continuing this practice and, over time, they have lost this opportunity of income generation altogether.

6.2.2 Vulnerability in household assets

The vulnerability of physical assets of households is examined in this section. These include: houses, furniture, agricultural machinery, radios, televisions, sewing machines, bicycles, livestock, poultry, etc. Women in the study area were in general poor and so had few assets to mention. Those they did included: a few goats, few chickens or ducks, a small amount of jewellery and some pieces of furniture. Whatever they owned, however, was thought of as very valuable in respect to the outcome of various climatic events. In describing the condition of household assets, a respondent Bijli Rani from Sonakhali village stated that:

> I have little furniture in my house including a radio. Whatever we have are very precious to us, because if cyclones or storm surges attack, we would lose all of these and may not be able to buy agaom. I got few gold jewellery during my marriage, but after Aila I had to sell those to cope with the disaster effects.

The destruction of houses and assets by disasters such as cyclones, floods or water surge is a common impact in this area. During the catastrophe associated with Cyclone Aila, most study participants lost the majority of their assets, including livestock and poultry. This consequently made their overall livelihood more vulnerable. About 90 per cent of women in the study responded that their assets were severely vulnerable in respect to climate change effects (Table 6.1) and that it was very difficult for them to recover from this situation. There was no household that thought that there was no vulnerability posed to their asset ownership as a result of climate change effects.

6.2.3 Vulnerability of lives and health

Women's lives and health are more vulnerable than men in any time of climate variability. More women die during natural disasters compared to men because they are not adequately warned, cannot swim well or cannot leave the house alone (UNFCCC, 2005). During Cyclone Aila in 2009, the death rate of women was counted as being five times higher than that of men (UN, 2010). In a cyclone, as suggested previously, even if a warning is issued, many women die while waiting for their relatives to return home and accompany them to a safe place. They cannot move without the concern and help of the male household head. Moreover, the majority of women do not know how to swim. Also, the kind of clothes women wear restrict their mobility, and having to protect their children can make a quick escape during an emergency situation difficult (Dasgupta et al., 2010). Poorly constructed and insecure housing systems also increase the mortality rate of women as they simply collapse during cyclone events. Women also suffer from various diseases and injuries during and after times of disaster. Pregnant women, lactating mothers and differently abled (disabled) women suffer the most as they find it difficult to quickly move to safety before and after disasters hit (WEDO, 2008).

The respondents were asked whether they or any of their female family members had suffered any major health problems from the past climatic events, and 60 per cent of women said that they had faced problems. Moreover, they were continuously suffering from different health-related issues. Among the respondents, 90 per cent of women said they were facing severe to medium levels of vulnerability in regards to health issues (Table 6.1). In response to this point, a key informant-10 revealed that:

> In our area women and adolescence girls are affected by gynecological and dermatological problems by using saline water in all their household works. Often they feel shy to inform this problem to other members of the family. Eventually, the situation gets worse and they cannot bear the expenses for treatment.

The FGD participants revealed that their health conditions had deteriorated in recent years. They spoke of suffering from frequent and various diseases including diarrhoea, dysentery, cold and flu and skin diseases. The reasons for this were suggested to be associated with: excessive heat and rain, saline water, water logging and a lack of nutritious food. As women have to do most of the water-related work such as collecting drinking water, cooking, washing, bathing the children, cleaning kitchen utensils and so on, they therefore, were the ones who were mostly affected by and at risk from water borne diseases. Excess salinity in water and the water logging situation all the year round also created various health problems for women in the study area. In general, a lack of medical facilities, malnutrition, lack of pure drinking water and proper sanitation facilitates all impacted upon and contributed to the vulnerability of women's health.

6.2.4 Vulnerability of food security

In the context of climate change, traditional food sources become more unpredictable and scarce, and this has had serious ramifications on four dimensions of food security: food availability, food accessibility, food utilisation and food systems stability (UN WomenWatch, 2009). It was found that the food situation was very dire for women in the study area, particularly those from the lower economic groups, and 70 per cent of respondents subscribed to this view. Another 13 per cent of respondents thought that their vulnerability was at a medium or moderate level. One of the main reasons was that due to salinity, the crop production in that area had significantly reduced. Among the respondents, 90 per cent of the women said that their land was degraded due to salinity and it had lost its capacity to grow crops. This had serious implications for their ongoing food security. However, 6 per cent of respondents who were in a higher economic group stated that they had not felt any vulnerability in regards to food security due to climate change events (Table 6.1). A key informant-5 stated that:

> Due to our social norm, women are usually the last to eat and receive the least amount in the family. This can often mean that poor women may pass several days without any food or less amount of food. As a result, they became more vulnerable to the unavailability of food and associated malnutrition.

The FGD participants and key informants stated that during Cyclone Aila, many women took shelter in the river embankments where there was no facility to cook, and more importantly they did not have money to buy food. They were eagerly waiting and wondering when the speedboat with food from the government and NGOs would come and give them some food, mostly dry food. During that time, families had to live on dry food for more than two months. They manage to eat only once a day, and that always consisted of the same dry food, such as flattened rice, biscuits or puffed rice. In addition, the amount of food they received was not sufficient for the whole family; thus, in many cases women would usually sacrifice their rations to feed other members of the family first. As a result of this prolonged food deficiency, most of the women in the study area were suffering from malnutrition. Natural disasters had also destroyed their crops, vegetable gardens and livestock and poultry (a major source for their daily food intake). Poor women were often forced to collect wild leaves, herbs and taro as food with or without rice. Many days they had to fast as there was simply no food left for them. Overall the food security situation was dire for the majority of the poor women and more or less a concern for all.

6.2.5 Vulnerability of education

The term 'vulnerability of education' is used here to represent whether women in the study area faced any vulnerability in providing education for their children; this is included as a part of their overall livelihood strategies. While the respondents themselves did not have much opportunity to be educated, their children were facing a different scenario. The respondents thought that their children were generally not vulnerable to having a lack of access to education due to climate change. Although there were some interruptions in education in post-disaster times, overall they were happy with the educational services they were receiving for their children. Therefore, most of the respondents (87 per cent) thought that there was no vulnerability in the case of education (Table 6.1). At present, the numbers of primary schools in Padmapukur and Gabura unions are 14 and 12 respectively (Bangladesh National Portal, 2015). While discussing the education facility in the study area, a key informant-8 stated that:

> Now government is emphasizing on education and set target to reduce the drop out rate at primary level. In recent times, many primary schools have been established in the locality where children can study free of cost. Therefore the attendance rate in primary school is very high. In addition, the social and cultural barriers for girls' education also have been reduced by the help of government and NGOs. Now, most of the children can go to school at least to primary level.

A few respondents (13 per cent), however, did think that there were vulnerabilities in relation to education to some extent as they were faced with problems of children dropping out due to the increasing demand for labour in households, from those experiencing difficulty in getting to school due to poor transportation systems after disasters, and the overall destruction of education materials during these events.

6.2.6 Vulnerability in relation to water sources

In rural areas all over the developing world, women and girls bear the burden of fetching water for their families and spend significant amounts of time daily hauling water from distant sources, even though the water from distant sources is rarely enough to meet the needs of the household (UN Women-Watch, 2009). The FGD participants said that water supply had been a problem in the study area for a long time. People were dependent mostly on surface water bodies (e.g. ponds). Cyclone Aila brought about saline intrusions into and over surface water resources, thus making them unfit for drinking and other uses.

The level of vulnerability mainly depends on the availability of freshwater and the distance of water sources from the household. Since women were responsible for most of the water-related works, they faced much more vulnerability than men due to a lack of local freshwater sources. Thus, women needed to travel long distances to collect water, often having to go to other places by crossing a river. This consumed an enormous amount of their time and they faced difficulties in completing other household chores. Sometimes in the rainy season, they were able to preserve rainwater and use it all year round for household purposes, though generally, the required amount of water for daily household jobs was never sufficient. Women in Gabura union in particular were more vulnerable than those of Padmapukur union, as the number of water sources was very limited in this union. Overall, 37 per cent of participants mentioned this being an issue of great concern. Conversely, 43 per cent of respondents were of the opinion that they were not vulnerable to water source problems as they were living nearby to water sources, provided by the government or through NGO support (Table 6.1).

6.2.7 *Vulnerability in relation to sanitation*

In Bangladesh natural disasters occur at regular intervals and a large number of sanitation facilities are destroyed every time. The FGD participants stated that after Aila they had to live in a tent on the riverbank where there were no toilets, and they had to share the same space with men. It was a very embarrassing situation for women as they had to go to common open places or temporary places for this purpose. Many women refrained from using the toilet during the day and consequently suffered from urinary tract infections and other diseases. For pregnant women, adolescent girls and elderly women, the situation was more difficult. Moreover, they faced problems in bathing and management of general hygiene aspects while they lived under the open sky with no private spaces. Overall, during the post-disaster time, the sanitation facilities were disastrous and women bore the brunt of this. However, after Cyclone Aila, different NGOs took the initiative to build toilets in many households and taught people the importance of the hygienic use of them. Therefore, women today are now in a better position than before in regards to sanitation facilities.

A respondent named Aloka Rani from Sonakhali village said that:

> Our house had no structured toilet before. It was very inconvenient for us. But after Aila NGO Caritas has made toilet in our house. Now, we are happy with our sanitation facility.

Ultimately, women who were the recipients of toilets from NGOs (78 per cent) did not record feeling vulnerable in regards to this, though they did face problems during the disasters. There remain, however, many households who still do not own a toilet (22 per cent) and the women of these households thought they were very vulnerable without this facility, with 13 per cent of respondents stating that they were severely vulnerable to sanitation problems (Table 6.1).

6.2.8 Vulnerability in relation to shelters and security during times of disaster

The overall capacity of cyclone centres was found to be very low and decreasing further with the increasing population numbers each passing year; this is despite the fact that climate change events in Bangladesh are predicted to increase exponentially (IPCC, 2007). There are only three cyclone centres in Padmapukur union and two in Gabura union (Bangladesh National Portal, 2015). These shelters differ from other public buildings in the sense that they serve a dual purpose: they are schools or health centres under normal conditions and refuges for people during and after cyclones (Khan, 2008). Typically, Multi-purpose Cyclone Centre (MPCS) facilities are multi-storied buildings with reinforced concrete that can, on average, accommodate up to 1600 people. These shelters generally have an open ground-floor structure to avoid flooding from the storm surges and the top floors (either one or two) are designed to accommodate people during and after the disasters (World Bank, 2010).

In the present study, 97 per cent of women stated that they felt vulnerable in these disaster shelters (Table 6.1). During FGD in Gabura village, a participant named Kulsum Begum stated that:

> Our houses are mostly built from earthen walls and golpata. During disaster time it is not safe to stay in these houses as it could collapse or wash away. We therefore try to leave our houses and go to cyclone centres or on river embankments during or after disaster time.

The cyclone shelters become full and even beyond capacity very quickly during cyclones. Often cyclone centres are far away and sometimes there is no cyclone centre within reach during disasters. Women often find it risky to go to the shelters if not accompanied by males. Elderly women and pregnant women find it even more difficult to walk along muddy roads under stormy conditions. Even when women are willing to take shelter in such centres, their husbands are often found to be reluctant to accompany them due to anticipated adverse situations.

During disasters (e.g. Aila), due to great distances from their households and inadequate space in the cyclone centres, many people took shelter in neighbouring embankments under open skies, which was very insecure for women, especially for young women. A UN Assessment Team found that nearly 14,000 families were living on the embankments, even one year after Cyclone Aila (UN, 2010). A respondent named Murshida from Jelekhali village of Gabura union described her experience of staying on the embankments during Aila as:

> There was bad smell all around us as dead bodies of various animals were floating in the water around us, and due to water logging these couldn't be removed. We had no place to go but stay there. Often I vomited by losing my tolerance to

that situation. There was water all around us but no water to use. There was no toilet and/or bathing facilities. That was a terrible time we passed.

The focus group discussants also stated that the shelters on the embankments were far below the minimum standard; they were mostly tents or one-room mobile huts without any space for privacy and were extremely vulnerable to the monsoon. However, without the repair of the damaged houses in the unions, people were unable to return to their houses. It took a long time to rebuild or repair their houses after the disaster and since most of their livelihoods had been destroyed during this period, it was beyond the people's capacity to rebuild their houses without the external assistance provided by government and NGOs. This situation also prolonged their stay at the temporary shelters and increased their vulnerability overall.

6.2.9 *Vulnerability in communication and transportation*

The means of communication can become extremely vulnerable due to climate change effects as seen in the study areas. Women of the study area (52 per cent) thought that they were 'severely' vulnerable to changes in communication and transportation due to climate change effects, whilst 38 per cent stated they felt 'moderately' vulnerable. Only 7 per cent of the participants registered being 'unaffected' by these events (Table 6.1). When Aila hit the study area, most of the roads were destroyed. Moreover, due to the heavy salinity, flooding and water logging problems, communication systems also collapsed in many areas. In the rainy season, during the heavy rainfall, the movement from one place to another was near impossible. Also, no transportation worked on those muddy roads, so it was difficult for women to collect water in this situation and for children to go to school. Many of the roads in the village areas are made of mud. Therefore, if any event occurs (e.g. cyclone, water surge, flood), those roads break down and people are essentially stranded.

A key informant-6 described the situation as:

In this area, communication system was not good even before Aila. But during Aila, all village roads were destroyed. At that time people had to depend on boats as a mode of transportation. Afterwards, with the help of government and NGOs, village roads were repaired and still the works are ongoing. However, as the roads are not paved, we suffer a lot in the rainy season.

Overall, women were found to be vulnerable in various aspects of their livelihood. If proper training and awareness programmes on disaster awareness and preparedness could be organised for them, their vulnerability might be somewhat reduced. The respondents in personal interviews were asked whether they had received any training on disaster preparedness or awareness from the government or by any organisation, and 100 per cent of the respondents stated that they 'never' received any sort of training. This kind of training could enable them to become familiar with the process of adaptation to climate change and its related effects more easily. However, thus far, women have mostly had to find their own strategies to cope within these stressed situations. The respondents outlined some of these coping strategies and they are discussed below.

6.3 Coping strategies of women against climatic events

Coping strategies refer to the specific efforts, both behavioural and psychological, that people employ to master, tolerate, reduce or minimise stressful events (MacArthur Research Network on Socioeconomic Status and Health, 2015). In disaster management, 'coping strategies' can be defined as a set of activities or mechanisms people try to utilise to survive during disasters and which, in turn, allow them to recover their situation and further develop their conditions after disaster (Islam, 2011). Since women are vulnerable in different ways to men under climate variability, they also have developed their own coping mechanisms. Some observed coping practices in relation to specific climate-induced vulnerability, as pointed out by women in personal interviews and FGDs, and by key informants of the study, are discussed in the section below. The frequency of observed coping strategies were broken down into four categories: 'always,' 'occasionally,' 'very rarely' and 'never.' The breakdown of percentages of women utilising specific coping strategies are presented in Table 6.2.

6.3.1 Reduced food intake in and after times of disaster: a common practice

As suggested earlier, availability of food is the main concern for the households in the study area, especially in times of disaster. When a household faces a food crisis during or after a disaster, women are responsible for adjusting household food consumption by either changing the type of food eaten, or by personally consuming less or nothing at all. All respondents (100 per cent) in the study area said

Table 6.2 Disaster coping mechanisms of women

Coping strategies	Frequency of practiced mechanism			
	Always	Occasionally	Very rarely	Never
Reduce food intake	60%	30%	10%	0%
Selling of assets	33%	49%	10%	8%
Receiving credit	35%	63%	2%	0%
Using saving	99%	1%	0%	0%
Alternative livelihoods	11%	84%	4%	1%
VGD card	69%	3%	1%	27%
Homestead gardening and poultry rearing	90%	6%	2%	2%

that they took less food during and after times of disaster to adapt to the changing scenario. Sixty per cent of women said that they 'always' practiced this self-imposed starvation, whereas 30 per cent practiced it 'occasionally,' and another 10 per cent practiced this 'very rarely' (Table 6.2). A respondent named Shefali from Patakhali village stated that:

> We used to produce crops before, and we had livestock and poultry. I also produced some vegetables and fruits in front of my house. We lost all these because of Aila and salinity problem. Now, we have to buy every food item which is expensive for us. Therefore, we eat less now.

It was also a difficult task for women to collect cooking utensils, ovens, fuel and food materials in and after disasters. For use during and after disaster times they preserved firewood, dry foods (such as rice, puffed rice, flattened rice and molasses), candles, matches and portable mud stoves in dry places, where and when they could. Despite the limited resources in the coastal areas, women played a significant role in food preservation to combat adverse situations.

6.3.2 Selling assets: the foremost strategy to adapt to adverse situations

The study found that the method that most applied in order to recover from the disadvantageous situation in which coastal communities of Bangladesh found themselves was the selling of their resources or assets (Paul & Routray, 2010). In the study area, about 33 per cent of respondents stated that they 'always' sold their assets during times of disaster, whereas about 49 per cent of respondents 'occasionally' sold assets then and about 10 per cent of respondents 'very rarely' sold their assets at all (Table 6.2). Women whose households had land attached to them sometimes mortgaged their land to others or tried to grow salinity-tolerant crops on their land. Few women in the study area, however, had their own land; therefore, selling other assets was considered the best alternative for those who found themselves in dire straits. In order to meet household financial needs during or after disasters, women usually sold any surviving livestock, poultry, stored crops, seeds, ornaments, trees or whatever they owned. The most popular objects to sell were livestock and poultry and not simply because of the urgent monetary needs, but also due to the lack of fodder and adequate shelter for storing ruminants. However, most did not receive an appropriate price for their assets, particularly by selling them during the disaster period rather than in normal times, as their desperation was evident to buyers. Many respondents mentioned, however, that they had no option but to sell their assets to cope with the disaster-driven situation, but at the same time, they were aware that they were having to sell at heavily discounted prices. Conversely, only 8 per cent of respondents 'never' sold assets, the main reason being that they did not have any assets that could be sold.

6.3.3 Receiving credit: a most demanding adaptation option

Receiving credit from informal sources is a practice dating back hundreds of years. Village people usually have seasonal income related to their crop harvesting. During other times of the year they take credit from village credit lenders to support their needs. Recently, institutional credit has been added to this arena, and its target is mainly women at the village level. Many women in rural areas are now part of microfinance organisations, using their memberships to access loans. The Grameen Bank is the pioneer institution which gives small credit to women. Women rely on credit from different sources, which is the easiest option for them to use to adapt when faced with disadvantaged situations.

Among the respondents, 35 per cent said that they 'always' took credit on all occasions during disasters, 63 per cent took it 'occasionally,' and another two per cent took credit 'very rarely' (Table 6.2). They used this credit for buying food and medicines, repairing houses, buying poultry, livestock, and/or boats or nets, children's education or marriage, or any other purposes. They said that without the credit facility, it would have been very difficult for them to adapt to the ongoing situation.

Prior to Cyclone Aila, women usually took credit from their relatives and neighbours for survival, as they felt more comfortable using this source. However, after Cyclone Aila, different national and international NGOs took strong initiatives to rescue and support the livelihood of the people of the study area, and supplying credit was a part of their actions.

6.3.4 Using savings: the firsthand option for adaptation

Although rural women in Bangladesh do not earn much and have limited income opportunities, from past generations they have learnt to save something for rainy days from their daily expenses either in kind or in cash. Often they save some rice every day from their daily uses or a little amount of money day by day. Most of the women also treat gold or silver jewellery which they received in their marriage ceremony as their savings. They also sell their ornaments in the lean periods. The total amount of savings may not be a significant amount to mention, but it is the first option for them to use to survive in times of disaster when their livelihoods have all but been destroyed. Not surprisingly, at that time, this little bit of saving meant a lot to them.

Almost all of the respondents expressed the same opinion that they used their savings during or after the disaster (Table 6.2). This was the quickest and most straightforward option for them to cope with the changing situation since it would have taken time to implement other coping mechanisms. However, this would not have been a permanent option for long-term adaptation. Considering their meagre amount of savings, they could only utilise them for a short period of time. However, as they immediately faced the challenges of basic needs and survival of their life, their savings at least helped them to survive until they managed to find an alternative option to help them cope and slow-adapt.

6.3.5 Alternative livelihoods: a sustainable adaptation option

Due to the geographical and socio-economic conditions present in the study area, livelihood options are limited for all people, but especially for women. The destruction of standing crops, fisheries, livestock and poultry, and other household assets has drastically reduced the livelihood earnings of women due to the series of climate-induced disasters. It has become very difficult, if not near impossible, for women to bring back these livelihood options post disaster. Cyclone Aila also had a drastic impact on employment options and earnings of day labourers, which constituted more than 80 per cent of the local labour market (Shamsuddoha et al., 2013). After Aila and other climate change impacts, the work opportunities for men were also dramatically reduced and, as a result, most of the men left their houses to search for jobs in other districts. They left their families behind for considerable periods of time, during which they transferred all their responsibilities onto the shoulders of their womenfolk.

To cope with this added burden, 84 per cent of respondents said that they tried to find alternative livelihoods 'occasionally' when they could, while 11 per cent of women, due to dire circumstances, were forced to do this more often (Table 6.2). Women often caught fish and shrimp fingerlings from the nearby rivers to sell to the local traders, though it was not deemed prestigious work for housewives. Sometimes they worked in wealthy people's houses, and collected forest and other naturally grown foods to feed their family. This was obviously a very distressing and demeaning situation for many women. Some women also had to start work under relief programmes. They were engaged in road construction work where they dug soil. However, they often undertook this work by going against the long-held cultural barriers, suggesting that starvation was seen as a more powerful force than disobeying cultural norms.

Some local NGOs (e.g. Shushilon, Uttoron, NGF, etc.) gave support to women by involving them in some income-generating activities, such as goat rearing, sewing, mat preparation, honey culture, crab fattening, etc. The livelihood situation of women was described by a key informant-1 by this way:

> Our NGO is working to improve the livelihood of women of this area. Women of this area were used to raise livestock and poultry, and earned some money. But after Aila they are not able to do this practice anymore. We are particularly working with rural women as they have now few or no income opportunities. They are not interested to work outside of their house. Therefore, we are giving them training on tailoring, making cap, making mat and net, so that they earn some money from house. We also provide them sewing machine and relevant materials to do these activities at home.

These alternative livelihood options, however, are not yet permanent for many and in the short run, while they may help some adapt to their situation, overall, broader-spectrum initiatives are needed that focus specifically on women such that they can bring about long-lasting and sustainable changes in their livelihoods.

6.3.6 Using a Vulnerable Group Development (VGD) card: a government initiative

The Vulnerable Group Development (VGD) programme of the Bangladesh government aims at ensuring the socio-economic development of the poverty stricken and destitute rural women of Bangladesh. The objectives of this programme are to overcome prevailing food insecurity, malnourishment, financial insecurity and social degradation. Between 2011 and 2013, under the umbrella of the VGD, each month 30 kilograms of rice or wheat were distributed among 750,000 poor women throughout the country for 24 months (MoF, 2015). A key informant-3 explained that:

> To help the poor rural women we are running VGD programme in Shyamnagar upazila. There are 3425 VGD cards being used by the women of Shyamnagar upazila. Padmapukur and Gabura are unprivileged areas of this upazila. We are conducting some projects to improve the livelihood of these areas such as Test Relief (TR), Food for Work (FFW), and Employment Generation Programme for the Poorest (EGPP). We are also running some projects to build roads and culverts.

Women in the study area received some benefits from these programmes which helped them to survive during and post disaster times. About 69 per cent of the respondents took up the opportunity of VGD cards 'always' to adapt their livelihoods during climate change affected scenarios (Table 6.2). However, 27 per cent of respondents 'never' used this facility, as they were not in vulnerable groups according to the local government's criteria.

6.3.7 Homestead gardening and poultry rearing: an option that women are comfortable with

In rural areas of Bangladesh, it is a common practice for women to raise poultry and practise homestead gardening to support their livelihoods in any stressed situation. The situation, as highlighted above, however, is different for the present study area compared to other parts of Bangladesh. Due to adverse climate change effects, it became difficult for women to succeed in homestead gardening and livestock and poultry rearing; however, many wanted to continue to persist with these activities if they could rather than venture outside the household to seek alternative income generation activities. As mentioned previously though, during Cyclone Aila, all the livestock and poultry were washed away and gardens were totally destroyed. Thus, the restoration of livestock rearing has been almost impossible since there was no grazing land left in the local vicinity. Poultry rearing was being practised to a smaller extent, but shortage of feed made this very problematic. The soil and water had also become totally salinised. As a result, women had no place to feed to raise poultry or grow vegetables. Despite their obviously disadvantaged situations, women still tried to raise poultry and grow vegetables where they could, tending to think that these were the works that they could do and were most familiar with, even though realistically, they offered no real chance of success in supporting their livelihoods. About 90 per cent of the respondents of the study said that they preferred to raise poultry and do vegetable gardening to cope with the worsening scenario rather than take on other work outside the homestead (Table 6.2).

The whole scenario discussed above portrays the poor coping mechanisms that are held by women in the study area. While women try their best to cope with the altered hydro-geophysical conditions, it is at a cost of personal health and wellbeing and involves great sacrifice. Women's coping efforts remain severely challenged and limited by gender relationships, which seem unwilling to bend or alter, even in times of great distress. A huge initiative is needed to improve the present condition of women by removing the barriers for them and helping them adapt to the ongoing changing impacts of climate change. This, however, can only come through stricter government policies, the continued work of NGOs and, further discussion with religious and local leaders.

6.4 Accessibility of women to major welfare facilities

The previous section revealed that women struggled to adapt to ongoing climate change scenarios. Accessibility to major welfare facilities is another indicator that can be used to assess the adaptation capacity of any target group to climate variability. Women's livelihoods and adaptation can be strengthened by ensuring women's access to, control and ownership of resources and access to development resources (credit, information, training), plus culturally appropriate technology (Nellemann, Verma, & Hislop, 2011). Therefore, it is very important to evaluate the degree of women's accessibility to different facilities that can support their resilience and livelihood. These accessibilities can truly make a difference to their lives and personal wellbeing. It was found from the discussion with FGD participants and KII that women of the study area have historically experienced underprivileged lives for a very long time and as such, prominently defined ideas of 'quality of life' and so on, as meaning little to them. If they can eat two meals a day, or have two sets of clothes to wear during their lifetime, they believe that they have all that is needed to survive. Owing to the remote location of the study area, the facilities normally required to meet the basic needs of people do not reach them. As such, their disadvantaged livelihoods are rarely noticed by the national government and international communities and the condition of women in these areas remains mostly unknown.

However, after the extreme climatic event of Aila, the situation changed. The devastating condition of the people suddenly came to the forefront of the national and international media's attention and so too the different levels of government and NGOs. They started work in the study area immediately after Aila hit there, and they are continuing their work to improve the very poor livelihoods of the people in the study area today. Women's accessibility to different welfare facilities is presented in Table 6.3. The accessibility levels are categorised as 'adequate access,' 'limited access' and 'no access' and evaluated according to perceptions of what this was like before (2009) and after (2013) Aila.

Table 6.3 Accessibility of women to major welfare facilities

Area of accessibility	Percentage of respondents					
	Adequate access		Limited access		No access	
	Before Aila	After Aila	Before Aila	After Aila	Before Aila	After Aila
Medical facilities	7	55	81	45	12	0
Family planning facilities	5	17	84	74	11	9
Banking	10	19	61	57	29	24
Local government organisation	20	25	30	35	50	40
Communication and transport	38	67	55	31	7	2
Electricity	5	8	3	1	92	91
Income opportunities	80	20	15	69	5	11

Table 6.3 reveals that before or prior to the effects of Cyclone Aila, women had limited or no access to medical facilities, including seeing doctors or receiving medicines. Even when they were in a critical condition while giving birth, there was no clinic, doctor or hospital to assist them. After Aila, some NGOs assisted them by providing medical facilities such that they were then able to visit a doctor periodically and acquire some free medicines for general illnesses. The proportion of women who received 'adequate' medical facilities increased to 55 per cent after Aila, from only 7 per cent before Aila. However, the family planning facilities were still not enough to help women, and little has changed since before Aila. Seventeen per cent of women had access to adequate family planning facilities in 2013, whilst it was only 5 per cent in 2009. While there have been some slight improvements in the accessibility to these services, for the majority of women, they remain inaccessible and much more needs to be done to ensure that women's needs are provided for on a broader scale.

Banking facilities have always been limited or unavailable to women in these areas. In the whole Shyamanagr upazila, there are branches of 13 nationalised and private commercial banks (BBS, 2013). Women in the study area were usually economically so poor that they did not have the capability to run a bank account. After Aila, the accessibility to banking facilities increased a little (19 per cent in 2013 from 10 per cent in 2009), but generally these were located far enough away to still make this facility unreachable. Similarly, women were not used to contact with local government organisations such as the Union Parishad, the Upazila Parishad including the Upazila Agricultural Office and the Upazila Fisheries Office, as men usually dealt with these authorities. The social and religious norms and values of the study area do not appreciate or facilitate the need for women to be mobile in respect of doing such jobs. Only 25 per cent of women had adequate access to these authorities, where needed.

The communication facilities such as roads and transport improved after Aila due to the direct involvement of government and non-government organisations.

Women (67 per cent) also felt more comfortable in using these improved facilities than before Aila (38 per cent). Nevertheless, it should be mentioned here that since a very poor communication system existed beforehand, having some stable, brick-made roads meant a lot to them. Overall, however, the transportation options are still limited and generally in very poor condition. Since most women have little idea about the world beyond their village, small changes appear to make them happy and satisfied. Again, when it comes to services like electricity, most of the respondents did not have any access to electricity (91 per cent). They often did not know that this was a basic facility that they deserved to have. Only a few people used solar power to produce some electricity in their houses and these were also supported by different NGOs.

Due to extreme and ongoing climate change effects, the most unfortunate thing that happened to the women of the study area was that they had lost many of their income opportunities (e.g. livestock and poultry production, homestead gardening, working in the crop field as a day labourer). Before Aila, 80 per cent of women had some income, whilst after Aila it had reduced to only 20 per cent (Table 6.3). Although some women were in search of alternative livelihoods, for most of them, the situation had become worse over time. Overall, women have always had to deal with these stresses, but increased climate change events have exacerbated them to the point where they appeared to be believing in some divine act or intervention, hoping that God would provide them with a better future. Again, without proper access to a different level of facilities, it is impossible for women to adapt to worsening situations and to grow resilience in all categories of their livelihoods.

Figure 6.1 represents the adequate accessibility of women to various welfare facilities before Aila (2009) and after Aila (2013), where the centre point of the

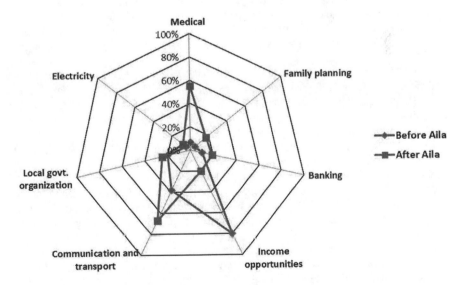

Figure 6.1 Accessibility of women to major welfare facilities in the study area

radar diagram represents zero access to facilities; the outer perimeter represents maximum access to facilities. It can be seen from Figure 6.1 that huge differences occurred in income opportunities in a negative way, and that over time, differences in medical, communication and transport facilities occurred in positive ways.

6.5 Conclusion

This chapter has explained that, due to their social and culturally constructed roles and responsibilities, and relatively poorer endowments of livelihood capitals, women were more vulnerable during and after climate-induced disasters such as Cyclone Alia. Therefore, access to land, more income opportunities and proper access to information and development facilities, increased independence and women's empowerment more generally must be ensured for the rural women of Bangladesh if they are to actively address climate vulnerability and strengthen their adaptive capacity for the future. Government and NGO support are much needed to facilitate these specific needs, as women still remain the most vulnerable group to climate change effects in the country. Having identified women's capital composition and the specific areas of vulnerability, the next chapter calculates two vulnerability indexes in order to quantify the degree of vulnerability of women's livelihoods.

References

Agarwal, B. (2003). Gender and land rights revisited: Exploring new prospects via the state, family and market. *Journal of Agrarian Change, 3*(1–2), 184–224.

Bangladesh National Portal. (2015). *Bangladesh national portal-information and service in a single window*. Retrieved from bangladesh.gov.bd

Baten, M. A., & Khan, N. A. (2010). *Gender issue in climate change discourse: Theory versus reality*. Dhaka: Unnayan Onneshan-The Innovators, Centre for Research and Action on Development.

BBS. (2013). *Literacy assessment survey (LAS)-2011*. Dhaka: Bangladesh Bureau of Statistics, Statistics and Informatics Division (SID), Ministry of Planning, Government of the People's Republic of Bangladesh.

Dasgupta, S., Siriner, I., & Sarathi De, P. (Eds.) (2010). *Women's encounter with disaster* (281 pp.). New Delhi, India: Frontpage Publications Ltd.

Hossain, M., & Tisdell, C. (2003). *Closing the gender gap in Bangladesh: Inequality in education, employment and earnings* (Working Paper No. 37). Social economics, policy and development series. Brisbane: School of Economics, University of Queensland.

IPCC. (2007). *Climate change 2007: Impacts, adaptation and vulnerability*. Contribution of Working Group II to the fourth assessment report of the Intergovernmental Panel on Climate Change (IPCC) (976 pp.). Cambridge and New York: Cambridge University Press.

Islam, R. (2011). Vulnerability and coping strategies of women in disaster: A study on coastal areas of Bangladesh. *The Arts Faculty Journal*, July 2010–June 2011, 157–169.

Jacobs, S. (2002). Land reform: Still a goal worth pursuing for rural women? *Journal of International Development, 14*(6), 887–898.

Kabeer, N. (2005) Gender equality and women's empowerment: A critical analysis of the Third Millennium Development Goal 1. *Gender & Development*, *13*(1), 13–24.

Khan, M. S. A. (2008). Disaster preparedness for sustainable development in Bangladesh. *Disaster Prevention and Management*, *17*(5), 662–671.

MacArthur Research Network on Socioeconomic Status and Health. (2015). *Coping strategies*. San Francisco: MacArthur Research Network on Socioeconomic Status and Health, University of California.

MoF. (2015). Empowering women and enhancing their social dignity, chapter 5: Ministry of women and children affairs. In *Gender budgeting report 2014–15*. Dhaka: Ministry of Finance (MoF), People's Republic of the Government of Bangladesh (GoB).

Nellemann, C., Verma, R., & Hislop, L. (Eds.) (2011). *Women at the frontline of climate change: Gender risks and hopes*. A Rapid Response Assessment of United Nations Environment Programme (UNEP), GRID, Arendal.

Paul, S. K., & Routray, J. K. (2010). Household response to cyclone and induced surge in coastal Bangladesh: Coping strategies and explanatory variables. *Natural Hazards*, *57*(2), 477–499.

Rahman, S. (2013). Climate change, disaster and gender vulnerability: A study on two divisions of Bangladesh. *American Journal of Human Ecology*, *2*(2), 72–82.

Shamsuddoha, M., Islam, M., Haque, M. A., Rahman, M. F., Roberts, E., Hasemann, A., & Roddick, S. (2013). *Local perspective on loss and damage in the context of extreme events: Insights from cyclone-affected communities in coastal Bangladesh*. Dhaka: Center for Participatory Research and Development (CRPD).

UN. (2010). *Cyclone Aila*. Joint UN multi-sector assessment and response framework. New York: United Nations.

UN WomenWatch. (2009). *Fact sheet: Women, gender equality and climate change*. United Nations Inter-Agency Network on Women and Gender Equity (IANWGE). Retrieved from www.un.org/womenwatch/feature/climate_change/

UNDP. (2010). *Gender, climate change and community-based adaptation*. New York: United Nations Development Programme (UNDP).

UNFCCC. (2005, December). *Global warming: Women matter*. United Nations Framework Convention on Climate Change (UNFCCC) COP Women's Statement, Montreal.

WEDO. (2008). *Gender, climate change and human security: Lessons from Bangladesh, Ghana and Senegal*. Report prepared by the Women's Environment and Development Organization (WEDO) with ABANTU for development in Ghana, action aid Bangladesh and ENDA in Senegal.

World Bank. (2010). *Vulnerability of Bangladesh to cyclones in a changing climate: Potential damages and adaptation cost* (Policy Research Working Paper 5280). Washington, DC: The World Bank.

7 The Livelihood Vulnerability Index

A pragmatic approach to measuring vulnerability

This chapter measures the vulnerability of women by the LVI and IPCC-LVI approaches (outlined in Chapter 3) based on the data collected from the household surveys and through the personal interviews with women. This chapter aims to calculate the level of livelihood vulnerability under the impacts of climate variability and extremes in the study area. This quantitative measurement of vulnerability is likely to provide greater gender-specific insights into the underlying processes and determinants of vulnerability, and is useful for translating the complex reality of vulnerability into simpler terms.

7.1 Components of livelihood capitals

The LVI used here was derived from 2013 data based on five livelihood capitals, that is: human, natural, financial, social and physical. Although these capitals and their related vulnerabilities have been discussed in previous chapters, this method will produce an individual index value for each of the capitals, which together will lead to the construction of a composite index for vulnerability, that is, LVI. Therefore, the selection of components and subcomponents under these five capitals in order to construct the indices was crucial. To proceed with the LVI, at first a vulnerability tree was developed to identify the major components and subcomponents of the five livelihood capitals (Figure 7.1) in the context of women's vulnerability in the study area. The major identified components under human capitals were: health, knowledge and skills, and livelihood strategies. The respective subcomponents under the components of health, knowledge and skills, and livelihood strategies are stated in Figure 7.1. Altogether, nine subcomponents were considered for measuring the vulnerability index of human capital, of which four belong to health, two belong to knowledge and skills, and three belong to livelihood strategies. As a contrast to these human capital elements, land, water, natural resources, natural disasters and climate variability (NDCV) have been identified as major components under the natural capital domain. In total, 14 subcomponents were considered under natural capital that belong to: land, water and natural resources, and NDCV components, and are presented in Figure 7.1.

In the case of financial capital, three major components were identified, namely annual income, asset ownership and finance. Overall, seven subcomponents were

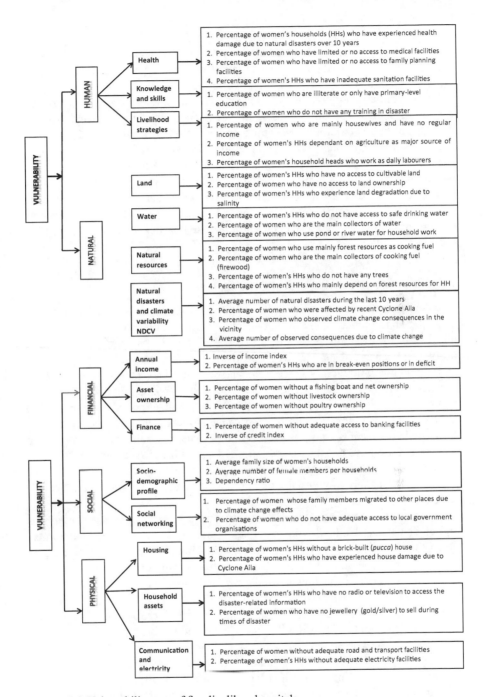

Figure 7.1 Vulnerability tree of five livelihood capitals

identified under financial capital according to: the annual income, finance, and asset ownership components. Social capital consists of two major components: socio-demographic profile and social networking. Five subcomponents (see Figure 7.1) were considered under this capital, of which three belong to socio-demographic profile and two under social networking.

Finally, three major components were identified under the category of physical capital, namely: housing, household assets, and communication and electricity. Each of these components was then divided into two further subcomponents, which resulted in a total of six subcomponents under physical capital (see Figure 7.1). Overall, 15 major components and 41 subcomponents were identified to measure the LVI for women in this study. The calculation procedures of major components and capitals were described in section in Chapter 3. The respective figures for the calculation of index numbers were derived from the outcomes of personal interviews with women and discussed earlier in various sections of Chapters 5 and 6.

7.2 Vulnerability index of human capital

In calculating the vulnerability index of human capital for women in the study area, the first component that was considered was health. The major contributing factors for deriving the health index were climate variability related health problems, the lack of proper sanitation facilities and a general lack of medical and family planning facilities. From the personal interviews, it was found that 60 per cent of women had experienced health-related problems as a result of several climatic events (see Table 7.1). In the study area, about 84 per cent of women did not have adequate sanitation facilities (no *pucca* toilets). Moreover, 83 per cent of women did not have adequate access to family planning facilities and 45 per cent women did not have adequate access to medical facilities (see Table 6.3). These figures altogether generated the health component index of 0.68, which indicates that the women were highly vulnerable to health-related issues as a result of climate change effects.

The second component that was considered for constructing human capital was knowledge and skills. Under this category, it was found from the personal interviews that women (100 per cent) in the study area 'never' received any training on climate preparedness or awareness (see Chapter 6). Moreover, about 73 per cent of the women's educational levels were very low (34 per cent women were illiterate, whereas 17 per cent could sign only and 22 per cent had only primary-level education) (see Chapter 5). These factors significantly increased their vulnerability. Thus, the vulnerability index for knowledge and skills was found to be 'very high' (0.865) for women.

Another vital component of human capital is livelihood strategies and the vulnerability index for this component was estimated at 0.737, which is also very high in terms of vulnerability. The reasons for this were that most of the women in the study (85 per cent) were housewives and spent most of their time engaged in household works, such as cooking, cleaning, taking care of the children and

Table 7.1 Vulnerability index of human capital

Component	Subcomponent	Unit	Observed value	Max value	Min value	Index value
Health	Percentage of women's HHs who had experienced health damage due to natural disasters over 10 years	%	60	100	0	0.6
	Percentage of women who had limited or no access to medical facilities	%	45	100	0	0.45
	Percentage of women who had limited or no access to family planning facilities	%	83	100	0	0.83
	Percentage of women's HHs who had inadequate sanitation facilities	%	84	100	0	0.84
						0.68
Knowledge and skills	Percentage of women who were illiterate or only have primary-level education	%	73	100	0	0.73
	Percentage of women who did not have any training on disaster preparedness or awareness	%	100	100	0	1
						0.865
Livelihood strategies	Percentage of women who were mainly housewives and had no regular income	%	85	100	0	0.85
	Percentage of women's HHs dependent on agriculture as major source of income	%	86	100	0	0.86
	Percentage of women's household heads who worked as day labourers	%	50	100	0	0.5
						0.737
Vulnerability index of human capital						**0.74**

elderly members of the family, fetching water, etc. (see Chapter 5). Again, since women were usually confined within their household area and not engaged in any income-generating activities outside, this certainly increased their livelihood vulnerability. Since, most of the women were housewives, their livelihood strategies mainly depended on their household heads, who were the main source of income for their families. In the present study, 86 per cent of women's households were solely dependent on agro-based livelihoods. This percentage represents the main sources of income for these households, which came from fish or shrimp farming (67 per cent) and crop production related agriculture (19 per cent), which can broadly be termed as part of the 'agricultural sector' (see Chapter 5). Since agriculture is very much dependent on climate, any climate variability affects this sector and, hence, the livelihood of their families. Therefore, dependency on agriculture increased the vulnerability of these women's livelihoods. Similarly, different climatic events reduced the work opportunities for the people in coastal areas. Therefore, women's household heads who worked as daily labourers (50 per cent) (see Chapter 5) were more vulnerable than any other occupational groups, which consequently increased the livelihood vulnerability of the women.

The vulnerability index for human capital was calculated by taking the weighted average of these three components (health, knowledge and skills, livelihood strategies), and was found to be 0.742, which can be termed as 'highly vulnerable.'

7.3 Vulnerability index of natural capital

Natural capital consists of four major components that include land, water, natural resources, natural disasters and climate variability (NDCV), and their respective vulnerability indices are presented in Table 7.2. When it comes to land resources, women in the study area were vulnerable at a measured level of 0.806, which is in the 'high' level. The main factor behind this was that 97 per cent of women did not have land ownership, which made their livelihood positions very vulnerable. Beside this, 55 per cent of women's households did not have access to cultivable land to grow any crops (see Chapter 5). Moreover, it was found from personal interviews that those who did have some amount of land faced a land degradation problem (90 per cent), mainly because of the salinity, which was a major climate-induced effect in the study area. Due to salinity problems, soil became unsuitable for growing any crops at all and the farmers were forced to sell or lease their land to wealthy shrimp farmers. At the end of the day, they ultimately found themselves as landless.

Water is another major component that affected the livelihood of women directly since they were the main collectors and users of water. The water vulnerability index was calculated at 0.62 for women of the study. Women were in a comparatively better position when utilising safe drinking water, as only 10 per cent of women used pond water and were therefore considered to be users of unsafe drinking water sources. Despite this, women still often had to travel a long way to collect drinking water which, overall, also affected their daily workloads and health. On average, 92 per cent of women were the main collectors of water

Table 7.2 Vulnerability index of natural capital

Component	Subcomponent	Unit	Observed value	Max value	Min value	Index value
Land	Percentage of women who had no access to land ownership	%	97	100	0	0.97
	Percentage of women's HHs who had no access to cultivable land	%	55	100	0	0.55
	Percentage of women's HHs who experienced land degradation due to soil salinity	%	90	100	0	0.90
						0.806
Water	Percentage of women's HHs who did not have access to safe drinking water	%	10	100	0	0.10
	Percentage of women who were the main collectors of water	%	92	100	0	0.92
	Percentage of women who used pond or river water for household work	%	84	100	0	0.84
						0.62
	Percentage of women's HHs who mainly depended on forest resources for HH income	%	27	100	0	0.27
Natural resources	Percentage of women who used mainly forest resources as cooking fuel	%	100	100	0	1
	Percentage of women who were the main collectors of cooking fuel (firewood)	%	5	100	0	0.05
	Percentage of women's HHs who did not have any trees	%	81	100	0	0.81
						0.533
Natural disasters and climate variability (NDCV)	Average number of natural disasters during the last 10 years	Number	6	7	4	0.67
	Percentage of women who were affected by recent Cyclone Aila	%	100	100	0	1
	Percentage of women who observed climate change	%	100	100	0	1
	Average number of observed consequences due to climate change in the vicinity	Number	4	6	1	0.60
						0.817
Vulnerability index of natural capital						**0.691**

in the study area. Since water collection is very labourious and time consuming work, women usually collected water only for drinking purposes. For other household purposes, about 84 per cent of women mostly used unsafe pond (81 per cent) or river (3 per cent) water (see Chapter 5). This practice also increased their livelihood vulnerability.

As suggested previously, the study area is adjacent to the Sundarbans. Therefore, the livelihoods of people of this area are heavily dependent upon forestry resources. Among the sample households, 27 per cent of households' income came from forestry resources (see Chapter 5). If they are constrained by any natural and institutional forces in collecting these resources, their livelihoods fall into risk. Almost all of the women (100 per cent) in the study area used forest wood for their cooking purposes. However, most mentioned that their husbands or male family members were the main collectors of this firewood. Only 5 per cent of women collected forest wood themselves to use as cooking fuel (see Chapter 5).

Another alarming natural constraint of the study area is the low number of trees in the whole vicinity. After the series of extreme climatic events, such as cyclones and tidal surges that occurred in recent times, most of the trees were destroyed. Also, as a prolonged consequence of climate variability, the soil became saline, most of the trees died and people failed to grow trees ever again. Thus, 81 per cent of the households had no trees in their household area or in other nearby places as only 19 per cent of the households had this asset (see Chapter 5). The combined effect of these factors resulted in a vulnerability index of 0.533 for natural resources.

The vulnerability index obtained for NDCV was 0.817 by the consideration of four subcomponents under this component. The respondents were asked about the number of extreme climatic events that they had experienced in the last 10 years. The responses varied from four to seven events that included: two cyclones, several tidal surges followed by seasonal storms, river erosion and floods. The average number of events recalled by the women was six. The recent Cyclone Aila had had a devastating effect in their lives, and all the sample women (100 per cent) had been affected by this catastrophe. The respondents were also asked whether they had experienced other examples of climate variability in their area. All of them (100 per cent) responded that they could recognise the changing climate in the locality. The respondent women also pointed out several consequences due to climate variability over the last 10 years. The responses varied from one to six consequences that included: increased salinity, excessive rainfall, increased temperature, extinction of trees, reduced fish stocks and increased diseases. On average they mentioned four climate-induced consequences in the area. Combining all the weighted values for these four components of natural capital (land, water, natural resources, NDCV), the total vulnerability index was estimated at 0.691 within 0 to 1 of the vulnerability range, and indicates that women in the study area were 'highly vulnerable' in terms of natural capital (Table 7.2).

7.4 Vulnerability index of financial capital

The vulnerability indices for components and subcomponents of financial capital are presented in Table 7.3. Income is one of the most important components

Table 7.3 Vulnerability index of financial capital

Component	Subcomponent	Unit	Observed value or index*	Max value	Min value	Index value
Annual income	Inverse of income index[1]	BDT	0.872 (0.147*)	1	0	0.872
	Percentage of women's HHs who were in break-even positions or in deficit	%	83	100	0	0.83
						0.851
Asset ownership	Percentage of women without a fishing boat and net ownership	%	98	100	0	0.98
	Percentage of women without livestock ownership	%	96	100	0	0.96
	Percentage of women without poultry ownership	%	64	100	0	0.64
						0.86
Finance	Percentage of women without adequate access to banking facility	%	81	100	0	0.81
	Inverse of credit index[2]	BDT	0.936 (0.068*)	1	0	0.936
						0.873
Vulnerability index of financial capital						**0.864**

* Observed income/credit index
1 Inverse of income index = $1/(1 + \text{income index})$
2 Inverse of credit index = $1/(1 + \text{credit index})$

of financial capital. The study considers the inverse of the income index as a subcomponent of annual income and was estimated at 0.872 which is 'highly vulnerable.' The average annual income for women's households was found to be at BDT 108,253 (US$ 1383), whereas the minimum and maximum annual incomes for women's households were found to be at BDT 20,500 (US$ 263) and BDT 616,800 (US$ 7906) respectively in the study area. Most of the households' incomes were closer to the minimum annual income. Considering their level of income, it is not surprising that 83 per cent of the respondents said that at by end of the year they were in deficit (13 per cent) or at a break-even position (70 per cent) in terms of income-expenditure ratio (see Chapter 5). Thus, the annual income index was estimated at 0.851. This scenario reflects the poor economic livelihoods of women in the study area.

When it comes to the asset ownership issue, women faced the worst-case scenario, and this was common for all rural women in Bangladesh. Many households in the study area would like to own a boat or net to catch fish from the nearby river, but the ownership of this fishing equipment, if it existed, solely belonged to men (98 per cent). In addition, most of the women (96 per cent) did not have any livestock ownership, though they were the caretakers of this asset, where it existed. Women also would have liked to raise some poultry in their homestead area, but here also, they could not enjoy the ownership right of this asset. Only 36 per cent of women were poultry owners in the study area (see Chapter 5). Overall, the asset ownership index for women was found to be at 0.86, which is undoubtedly 'high' in terms of vulnerability (Table 7.3).

Women also had limited connections to banks, as suggested previously, and 81 per cent of the women did not have adequate banking facilities. It should be mentioned here that there was no bank in the study area except in the upazila sadar (sub-district centre). It was difficult for women to get access to the bank within their present livelihood practices. However, most women did have some access to some local NGOs and had received small credit loans from them. They also used to receive credit from their relatives and some co-operative societies when they were in need. The average amount of credit received by women during the last five years was approximately BDT 20,500 (US$ 263), whereas the maximum amount of credit was found to be at BDT 300,000 (US$ 3846). However, many women had not taken any credit during last few years. Further information regarding this aspect of women's livelihood has been elaborated upon in Chapter 6. Combining these factors, the vulnerability index for finance components was obtained at 0.873, which obviously depicts the underprivileged financial facilities for women. Overall, the weighted values for annual income, asset ownership and finance components were responsible for the vulnerability index for financial capital being 0.864, which indicates 'high vulnerability.'

7.5 Vulnerability index of social capital

The two major components under social capital are socio-demographic profile and social networking. The socio-demographic profile of women mostly determined the vulnerability of social capital. The average family size of respondents' households was 5.05; the highest family size was found to be 12 persons and the lowest two persons. Since most of the female members of the family were not engaged in income-generating activities, it was assumed that the more female members in the family, the greater the risk of vulnerability that existed in the family. The average number of female members was found to be 2.44, whereas the maximum and minimum numbers of female members among the sample households were found to be six persons and one person respectively. The average dependency ratio of sample households was estimated at 0.664 and, hence, the index for this was calculated at 0.730. Considering these three subcomponents, the social-demographic index was registered at 0.441, which can be treated as being 'moderate' in terms of vulnerability.

Social networking was very limited for the women in the study area as they had very limited access to others outside their homesteads. Some of the male persons of the households (27 per cent) had migrated to other places in search of work, since work opportunities in the local area had decreased significantly due to the effects of climate change (see Chapter 5). During that time, women had to take

Table 7.4 Vulnerability index of social capital

Component	Subcomponent	Unit	Observed value	Max value	Min value	Index value
Socio- demographic profile	Average family size of women's HHs	Number	5.05	12	2	0.305
	Average number of female members per household	Number	2.44	6	1	0.288
	Dependency ratio	Ratio	0.664	0.91	0	0.730
						0.441
Social networking	Percentage of women whose family members migrated to other places due to climate change effects	%	27	100	0	0.27
	Percentage of women who did not have adequate access to local government organisations	%	75	100	0	0.75
						0.51
Vulnerability index of social capital						**0.469**

on the added responsibilities for the whole family and these created extra burdens upon them. Often, they found it difficult to manage their households and social responsibilities together. Women were also not used to contacting local government organisations such as the Upazila Parishad and the Union Parishad for any purpose, and 75 per cent of women neither had adequate access to these organisations nor were they familiar with the government services available through organisations (see Chapter 6). It was the male members of the family who usually went to see these organisations when they needed to. The social networking index was registered at 0.51. Overall, the weighted values of these two components (socio-demographic profile and social networking) resulted in a vulnerability index score of 0.469 for social capital, and the components of this index are presented in Table 7.4.

7.6 Vulnerability index of physical capital

The physical capital is naturally very vulnerable in relation to climate change effects. Any climatic event instantly affects the physical assets of any household. Therefore, the nature of physical capitals determines the vulnerability of this capital. The detailed vulnerability indices of physical capital are presented in Table 7.5. In terms of housing, those women who did not have brick-built (*pucca*) houses to live in were more vulnerable to climatic events. Unfortunately, this applied to 95 per cent of the sample women (see Chapter 5). The consequences of this were observed during Cyclone Aila in 2009. During the personal interviews, it

Table 7.5 Vulnerability index of physical capital

Component	Subcomponent	Unit	Observed value	Max value	Min value	Index value
Housing	Percentage of women's HHs without a brick-built (*pucca*) house	%	95	100	0	0.95
	Percentage of women's HHs who experienced house damage due to Cyclone Aila	%	97	100	0	0.97
						0.96
Household assets	Percentage of women's HHs who had no radio or television to access the disaster-related information	%	68	100	0	0.68
	Percentage of women who had no jewellery (gold/silver) to sell during times of disaster	%	98	100	0	0.98
						0.83
Communication and electricity	Percentage of women without adequate road and transport facilities	%	33	100	0	0.33
	Percentage of women's HHs without adequate electricity facilities	%	92	100	0	0.92
						0.625
Vulnerability index of physical capital						**0.805**

was found that 97 per cent of the households were severely damaged by this devastating cyclone. The housing index was registered at 0.96, which is 'extremely vulnerable.'

Other household assets, such as radios and televisions, can also reduce the risk of vulnerability. Households who have these assets can easily get the climate-related information or warnings, and take necessary actions. Moreover, government and international organisations broadcast many climate-related programmes through television and radio. This is most essential for the people of coastal communities. The outcomes of the personal interviewees revealed that 68 per cent of the women's household did not own either a television or a radio. This obviously increased their vulnerability to climate change event warnings. In addition, selling jewellery such as gold and silver in times of crisis was considered as one of the possible coping strategies for women, but 98 per cent of women revealed that they did not have much jewellery that could be sold in crisis situations (see Chapter 5). This also reduced their capacity to cope with the climatic variability and increased vulnerability. The vulnerability index for household assets was estimated at 0.83, which is 'highly vulnerable' (Table 7.5).

Adequate communication (e.g. roads, transport) and energy (e.g. electricity) facilities are other crucial indicators of physical capital status. Women in the study area were more or less satisfied with their present communication systems compared with a few years back. Still, 33 per cent of women thought that the road and transport facilities were not adequate for them (see Chapter 6). The most unfortunate thing for the women in the study area was that they did not have access to electricity facilities. This consequently made their livelihood harder as they were responsible for all household work, which was very labour intensive manual labour. Among the respondents, 92 per cent of women did not have adequate electricity facilities, as there were no electricity connections in Padmapukur and Gabura union (see Table 6.3). Only a few households received electricity through solar power. The index for communication and electricity was registered at 0.625. Taking into consideration the weighted values of these three components (housing, household assets, communication and electricity), the overall vulnerability index for physical capital was estimated at 0.805, and represents the 'highly vulnerable' situation of women in terms of physical capital.

Considering all the major components under five livelihood capitals, a vulnerability spider diagram is prepared and presented in Figure 7.2. The figure presents the vulnerability scores for 15 major components studied in the present research. The vulnerability scores for the major components ranging between 0 and 1, where 0 indicates 'no vulnerability' and 1 indicates 'highest vulnerability.' The further the score from the centre of the diagram, the more vulnerable are the components. The figure demonstrates that almost all of the livelihood components were 'highly vulnerable' for the women who lived in the climate change affected regions of the study area. Extrapolating from this, it can be inferred that the livelihoods of most

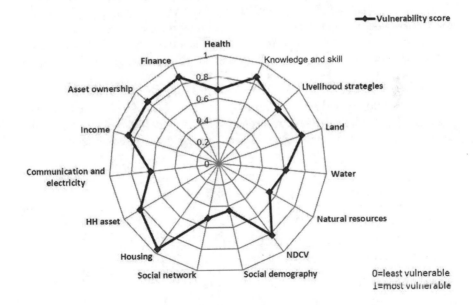

Figure 7.2 Vulnerability spider diagram of the major components of LVI for women in Shyamnagar upazila

women residing in the coastal areas of Bangladesh are at risk and will continue to remain so until concrete steps are taken to assist in their adaptation to the ongoing and adverse effects of climate change.

7.7 Composite Livelihood Vulnerability Index

The overall LVI was estimated based on the weighted vulnerability scores of the five livelihood capitals ranging from 0 to 1, and was found to be at 0.7286 for the women in the study (Table 7.6). Based on that result, it can be said that the livelihoods of rural women in the study area are 'highly vulnerable.' They are victimised by the composite effects of climate change and the poor socio-economic conditions of the study area as a whole, not to mention the continued extreme levels of patriarchy in their society. Thus, they can be recognised as the most vulnerable group in Bangladesh in the context of climate change effects. This result gives an indication of which capital should be taken into account for reducing livelihood vulnerability of women in the study area. Figure 7.3 represents the vulnerability indices for the five livelihood capitals, namely: human, natural, financial, social and physical capital. The figure shows that women were 'highly vulnerable' in terms of all of the capitals, particularly in the cases of financial and physical capital.

7.8 Presentation of the IPCC-LVI

The IPCC (2007b) defines livelihood vulnerability as a function of system exposure, sensitivity and adaptive capacity. The IPCC-LVI was computed by grouping the 15 major components that have been identified for LVI calculation into three categories, namely: exposure, sensitivity and adaptive capacity. The respective components under exposure, sensitivity and adaptive capacity, and their respective vulnerability indices are presented in Table 7.7.

Whereas exposure was made up of the score for only one major component (NDCV), sensitivity and adaptive capacity were made up of aggregated scores of eight (health, land, water, natural resources, housing, household assets, communication and electricity, and annual income) and six (socio-demographic profile,

Table 7.6 Livelihood vulnerability index of women in Shyamnagar upazila

Capital	Number of components	Vulnerability index by capital	Livelihood vulnerability index
Human capital (HC)	3	0.740	
Natural capital (NC)	4	0.691	**0.7286***
Financial capital (FC)	3	0.864	
Social capital (SC)	2	0.469	
Physical capital (PC)	3	0.805	

* Livelihood Vulnerability Index is a weighted average of five capital indices.

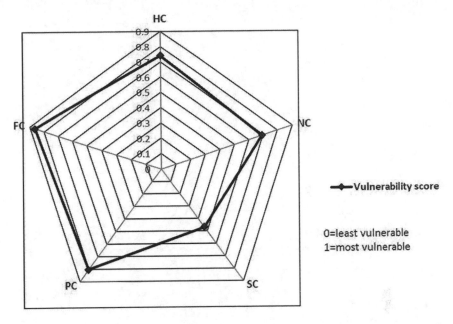

Figure 7.3 Vulnerability spider diagram of five capitals for women in Shyamnagar upazila

Table 7.7 Livelihood Vulnerability Index of women in Shyamnagar upazila by IPCC-LVI

IPCC contributing factors to vulnerability	Major component	Index value
Exposure	Natural disaster and climate variability (NDCV)	0.817
Exposure Index (EI)		**0.817**
Adaptive capacity	Socio-demographic profile	0.441
	Social network	0.51
	Livelihood strategies	0.737
	Asset ownership	0.86
	Finance	0.873
	Knowledge and skills	0.865
Adaptive Capacity Index (ACI)		**0.707**
Sensitivity	Health	0.68
	Land	0.806
	Water	0.62
	Natural resources	0.533
	Housing	0.96
	Household assets	0.83
	Communication and electricity	0.625
	Annual income	0.861
Sensitivity Index (SI)		**0.713**
IPCC-LVI = (EI – ACI)* SI		**0.078**

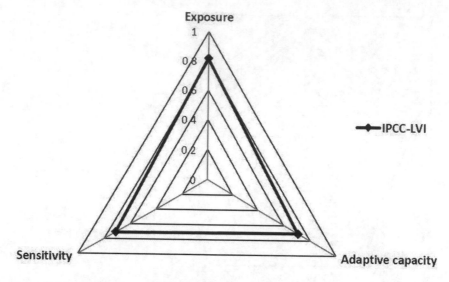

Figure 7.4 Vulnerability triangle of the contributing factors of the IPCC-LVI for women in Shyamnagar upazila

social networking, livelihood strategies, asset ownership, finance, and knowledge and skills) major components respectively. Under the consideration of similar indicators calculated by their respective methods, LVI and IPCC-LVI yielded consistent results in the present study. The aggregated scores of exposure, sensitivity and adaptive capacity are represented in the vulnerability triangle, as shown in Figure 7.4. The vulnerability triangle indicates that women were very much exposed (0.817) to the impacts of climate change and variability. Their livelihood was also very sensitive (0.713) in the context of several climatic effects. The women were also found to be vulnerable in terms of their overall adaptive capacity (0.707).

The IPCC-LVI is scaled from −1 (least vulnerable) to 1 (most vulnerable). The IPCC-LVI becoming positive means that the target group is more exposed to climate extremes and natural disasters than its capacity to adapt to or overcome those adverse situations. The function IPCC-LVI = (Exposure-Adaptive capacity) x Sensitivity, produced the IPCC-LVI score for women in the present study and it was recorded at 0.078. This value indicates a 'high level' of vulnerability to climate variations and changes as it is a positive value. It can be argued from the result that having a high exposure and a relatively low adaptive capacity causes the women of the study area to be highly sensitive to climate change and that these changes affect them in a negative manner. This index represents a linear relationship among the three contributing factors of vulnerability denoting women's high exposure and sensitiveness to climate change and variability, accompanied by a lower adaptive capacity.

7.9 Conclusion

This chapter has presented two alternative methods for measuring the vulnerability of livelihoods of women in a climate-susceptible scenario. The estimates provide a detailed depiction of factors driving the livelihood vulnerability of a special group (i.e. women) in a particular setting (e.g. affected by climate change). In both of the estimates, the livelihoods of women were found to be highly vulnerable. The results indicate that it is extremely important to instigate strategies to build the adaptive capacity of women to reduce their livelihood vulnerability. Only by doing this will they be able to face the ongoing risk of natural disasters and the increasing effects of climate change into the future.

In Bangladesh, this is the first-ever estimation of women's livelihood vulnerability using LVI and IPCC-LVI. This quantification of vulnerability is a significant contribution to the literature on climate change impacts on livelihood. These findings have important policy relevance for all involved in disaster and risk management, both within Bangladesh itself and outside. In developing countries in particular, this approach can facilitate the accurate assessment of livelihood of different groups of people and, thereby, allow for relevant strategies to put in place to address vulnerability and reduce disaster risks. Based on all the results presented thus far, the next chapter draws a conclusion to this book.

Reference

IPCC (2007b). *Climate change 2007: Impacts, adaptation and vulnerability. contribution of Working Group II to the Fourth Assessment Report of the Intergovernmental Panel on Climate Change (IPCC)* (976 pp.). Cambridge: Cambridge University Press, Cambridge.

8 Conclusion

8.1 Summary of contribution

This research has examined the livelihood capitals and vulnerability of women in the disaster-prone coastal areas of Bangladesh in relation to climate variability and climate change impacts. The coastal upazila of Shyamnagar was selected as the study area since it had experienced a number of natural disasters over the last few years. The study combined an SLF, a vulnerability assessment approach (Disaster Crunch Model) and a composite index approach as its methodology to develop the research framework and guide the overall data collection and analysis processes. This integrated approach enabled the investigation of local scale processes that influence livelihood vulnerability, while embedding these within the broader processes associated with climate change.

Following a mixed-method approach, the data was collected using household questionnaires, key informant interviews, focus group discussions, a transect walk, as well as secondary sources. The use of a mixed-method approach throughout the research enabled rigour to be maintained through the triangulation of data sources, while still permitting flexibility in data collection and the gathering of rich in-depth data to deepen the understanding of the issues raised by this research.

Chapter 4 outlined the geographical and socio-economic characteristics of the study area which facilitated a greater understanding of the livelihood setting of people in the coastal areas of Bangladesh. These areas, in many respects, were quite different and unfamiliar to the typical rural livelihoods experienced in the rest of the country. The researchers' own observations and previous experiences supported this fact. The detailed descriptions of the effects of Cylone Aila (perceived and otherwise) and the role of the Sundarbans in the maintenance of many livelihoods have provided a more nuanced level of understanding and specificity needed in the continued study of situated climate change effects.

Chapter 5 presented the outcomes of the livelihood capitals assessment based on an SLF approach. Five livelihood capitals – human, natural, financial, social and physical – were examined to assess the livelihoods of women within a setting of climate variability. The study compared women's livelihoods before and after Cyclone Aila to identify the changes in livelihood patterns. Overall, it was found that household incomes decreased after Aila, especially the income gained from agriculture, and that livelihood expenses had increased slightly. Thus, participants in this study were comparatively worse off after the devastating cyclone. Despite this, few people (3 per cent) have permanently migrated to other places. The amount of resources, such as livestock, poultry, fishing boats and nets, and trees belonging to the households have also decreased significantly due to the effect of Cyclone Aila. Overall, the distribution of the five livelihood capitals indicates a worsening livelihood structure for women when examined from the point of ongoing climate change effects.

Chapter 6 highlighted the areas of livelihood vulnerabilities of women to climate variability and extremes, that is, how the income, assets, health, food security, water sources, sanitation, social security, mobility and shelter were all vulnerable due to ongoing climatic hazards, such as cyclones and floods, salinity and water logging. In most of the cases, the vulnerability level was identified as 'severe' for the majority of respondents. The usual coping mechanisms practised by women and the frequency of the use of these techniques were described to identify and analyse their effectiveness in terms of long-term adaptive capacity. Adaptive capacity was also examined through the degree of accessibility to major welfare facilities, assuming that accessibility to these facilities strengthened adaptive capacity. Although there were some improvements in terms of accessibility, there remains a need for initiatives, both national and international, to help the adaptive capacity of women materialise so that resilience towards climate change effects might be learnt.

While Chapter 6 described vulnerability in a qualitative manner, Chapter 7 presented a quantitative approach to measure vulnerability. The composite index approach can be useful at the household scale to distinguish the factors that determine women's differential levels of vulnerability, and can be used to determine the prioritisation of factors required for the reduction of such vulnerability. The composite index approach to assess livelihood vulnerability has also contributed to a methodological discussion of approaches to vulnerability analysis more generally, particularly to the weighing of vulnerability indicators, validation of vulnerability indices and sub-indices, selection of significant vulnerability indicators, and an explanation of the role of indicators and sub-indices of vulnerability. The value of the LVI and IPCC-LVI indicates the vulnerability level of livelihood and quantifies how the different components and subcomponents affect that score. The results from the utilisation of both the LVI and IPCC-LVI reveal that the livelihoods of women are highly vulnerable in the rural coastal areas of Bangladesh.

Based on the findings of this study, it can be concluded that women in the coastal areas of Bangladesh have been adversely affected by numerous climatic shocks and stresses upon them in recent years. While they have traditionally coped with or adapted to the normal range of climate impacts, overall, in the face of increased climatic disasters, they have not coped well. Their livelihood vulnerability level remains at an alarming rate, and this may substantially increase in the coming decades because of ongoing climate change. Almost all of their livelihood capitals and the current coping strategies in place will be extremely tested in the face of increasing impacts from sea level rise, land erosion, cyclones and associated flooding, as these are predicted to be exacerbated due to global warming (IPCC, 2007b). The autonomous adaptive capacities they are currently using are not sufficient for them to address the present level of climate variability and change, let alone future predicted levels. Without appropriate and sufficient adaptation strategies, increased levels of climate change shocks and stresses will result in greater loss of women's lives and livelihoods in the coastal areas. In short, considering the extremely high level of vulnerability of women's livelihoods, it can be said that they are already in severe danger, and this will be further exacerbated in the future following predicted levels of climate change. Based on this investigation, the following broad recommendations for improving sustainable livelihoods of women in the coastal communities of Bangladesh are provided:

- Gender issues need to be identified and incorporated by the government into all planning stages and strategic initiatives against climate change. A gender mainstreaming plan that includes a detailed plan of action, a timetable, specific goals and accountability measures needs to be developed.
- Government and NGOs should promote women's empowerment through capacity building before, during and after climate-related disasters, as well as their active involvement in disaster anticipation, early warning and prevention programmes as part of their resilience building.
- Women's social, economic, physical and psychological vulnerabilities in community-based preparedness and response plans need to be acknowledged in order to reduce the impact of disasters on women.
- Community-based awareness raising programmes should be strengthened. Seminars, symposiums and workshops need to be arranged in coastal areas and the participation of women in these programmes need to be ensured.
- Women's access to land and assets, information and education, participation in decision making and self-dependence need to be increased through community development programmes. This needs to be a systemic approach and may require greater involvement of moderate religious clerics to ensure the message does not encounter unnecessary cultural bias.
- Government and NGOs should create opportunities for the creation of income-generation activities for women. It should be ensured that women have equal access to developmental facilities. Credit flow needs to be strengthened to support women's livelihoods.
- Salinity-tolerant crops, vegetables and fruits species should be introduced and disseminated to increase the opportunities for alternative livelihoods of women.
- Increased accessibility to drinking water needs to be provided. Facilities should be provided in order to preserve sufficient amounts of drinking water all the year round. Projects should also be taken to reduce salinity problems. Solutions may be found through international examples of best practice in other countries.

- The number of community health clinics should also be increased. The availability of doctors and nurses in health clinics should be ensured and monitored; this will ensure that the impacts from climate-induced events will be attended to immediately, rather than not at all, or months after they have been allowed to worsen. Basic medicines should be available to all women, irrespective of their level of income. This is a basic human right.
- New cyclone centres need to be built, gender-specific if need be, to ensure women's safety. Disaster shelters should be well constructed with multi-purpose uses and there should be separate facilities for women. Shelters should be constructed at places where access for women will be direct and easy during times of disaster.
- Infrastructure, such as roads and transport networks, should be further developed to expedite the movement of people. Electricity facilities should be provided in all coastal households, even at a basic level. If this cannot be done, then access to solar panels should be made available to all.
- Existing coastal embankments should also be repaired and maintained on a regular basis to reduce further disaster risks.

8.2 Research implications

Gender issues were absent for more than 20 years from the United Nations Framework Convention on Climate Change (UNFCCC) and largely from its associated decision-making bodies. The theme of gender and climate change was largely nonexistent on the global stage, and also in the National Adaptation Plan of Action (NAPAs) and national communications in Bangladesh itself. In 2009, the government of Bangladesh passed a climate change strategy and action plan where gender issues were, for the first time, addressed. Ongoing challenges remain, however, in the implementation of gender-oriented programmes within the context of climate change. The linkages between gender and climate change are less intuitive than one might expect, particularly in a country where a high level of patriarchy is the norm. This constraint could be one of the factors why the mainstreaming of gender in these areas generally has been more limited and challenging than most (MoEF, 2008).

The findings of this study are of particular relevance to the government of Bangladesh's Climate Change Strategy and Action Plan (BCCSAP). The research outcomes can provide a significant contribution to the BCCSAP's priority of gender-sensitive assessments and research, including collection, analysis and reporting of sex-disaggregated data to better understand the implications of climate change and climate policies. The study results allow for the identification of where greater understanding of the importance of gender and climate change are needed, and pinpointing possible areas of intervention for a specific action plan on gender and climate change in Bangladesh.

The findings of this study also allow the identification of a range of measures that could be utilised to help address the impacts of current and future climate variability and change in regards to women's livelihoods, particularly in the poorer, rurally based coastal communities of Bangladesh and, potentially, beyond. The women in the coastal areas will require multifaceted measures to be taken to ensure future adaptability and resilience. Global climate change mitigation is essential over the longer term to reduce exposure, overcome the limits to adaptation and build resilience, because adaptive capacity may be limited to only lower levels of climate change (≤2–3°C), and Bangladesh is slated to be at the forefront of much higher levels of climate change impacts (IPCC, 2007a). Much needed investment in these areas would clearly have an impact on those who have no choice but to live in these coastal areas. Investment in natural capital and the conservation of ecosystems can help to avoid crises and catastrophes or to soften and mitigate their consequences. However, given Bangladesh's limited economic resources, investment in hard infrastructure is unlikely in the near future, without external assistance. Ensuring improved livelihood outcomes for women by augmenting their livelihood assets and improving access to external sources will help to diversify livelihood strategies and provide other opportunities, which can only increase their overall wellbeing and reduce the daily stresses upon them.

Reducing the gender gap in social, political and institutional structures will all be required to effectively reduce the vulnerability of women's livelihoods in Bangladesh. Building human capital, such as through investment in education and skills development, could particularly help to diversify livelihoods, which in turn would help women and their households to become less reliant on activities that are climate-sensitive. Access to less expensive financial credit from government and non-government institutions would also ensure increased ownership of household assets, and increased income opportunities could help them in strengthening their financial and physical capital. Capacity development and the involvement of women in alternative technologies, for example through the use of bio-fertilizer, climate-resilient cropping, saline-resistant vegetables, fisheries culture and management, homestead gardening, technology for micronutrient-rich food, homestead plant nurseries and handicrafts training can all potentially increase the income opportunities for women. Essentially, in facilitating these opportunities, the government of Bangladesh will not only improve the livelihood of its most vulnerable citizens but also ensure long-term benefits for the country as a whole.

In addition, facilitating necessary migration, the proper management of forestry resources, adequate water and sanitation facilities and communication facilities could help them to enrich their natural and social capital endowments. Specific credit lines for women only and updating existing health policy frameworks to include gender and climate change linkages can be developed. In addition, developing the capacity of women to fully engage in water resource management committees, advocacy and awareness at central and local (community) levels, including engaging with print and electronic media and community radio for dissemination, should be ensured. These actions combined could help them to adapt to ongoing climate variability and shocks.

8.3 Future directions

To draw together the detailed insights gained into the vulnerability of women's livelihood in a climate susceptible setting, this research has focused on the local scale of one upazila. In order to draw general lessons more confidently by scaling up conclusions on livelihood vulnerability and adaptation to climate variability and change, a broader range of in-depth case studies, complemented by finer-scale climate and livelihood data, will be required. On the regional and global scales, climate change predictions suffer from important knowledge gaps and uncertainties, which need to be overcome. Wider-scale (district and country) studies on these issues may help generalise the findings. Cross-nation comparative case studies may also help countries learn from each other. Cross-country studies are a useful complement to single-country case studies, mainly because they provide information on variables, which may exhibit greater variation between countries than within them. Also, cross-country analyses can provide a basis for establishing policy priorities on a regional and global basis (IFPRI, 2000). Research initiatives can be taken to look at how the implications suggested in this study fit into the wider development arena.

Considering a very limited number of studies on gender and climate change currently exist, however, this study is a significant addition to the current literature on climate change in Bangladesh. No particular research has thus far been conducted on the vulnerability of women's livelihood in a disaster-susceptible setting such as Bangladesh. In this respect, it is a unique research project. In addition, the research has applied the LVI and IPCC-LVI approaches to measure the livelihood vulnerability of women. Initial scoping studies suggest that there has been no research in Bangladesh that has used this index approach to measure women's livelihood vulnerability in any setting. Therefore, this study also contributes to the methodological aspects of livelihood assessment. Overall, the findings of the study can provide valuable information to government and other policy makers to improve the livelihood of women in the coastal areas of Bangladesh, which are now further marginalised due to the impacts of climate change.

To fully utilise these contributions, additional research needs to be done on policy instruments of climate change that have yet to recognise the gender-specific characteristics of vulnerability and adaptive capacity. The role of institutions and civil societies can also be evaluated, as they play an important role in strengthening gender responsiveness to climate change. More climate change research that ensures the inclusion of gender-specific approaches may also balance out the currently inequitable distribution of climate change impacts. This could also potentially improve adaptive decision making, reduce negative impacts on the entire community and enhance human security overall. Ultimately, these findings are important not only for women in Bangladesh, but elsewhere where climate change impacts can be associated with entrenched inequality.

References

IFPRI. (2000). *Overcoming child malnutrition in developing countries – Past achievements and future choices. 2020 vision for food, agriculture, and the environment* (Discussion Paper 30, 73 pp.). Washington, DC: International Food Policy Research Institute (IFPRI).

IPCC. (2007a). *Climate change 2007: The physical science basis.* Contribution of Working Group I to the fourth assessment report of the Intergovernmental Panel on Climate Change IPCC) (996 pp.). Cambridge and New York: Cambridge University Press.

IPCC. (2007b). *Climate change 2007: Impacts, adaptation and vulnerability.* Contribution of Working Group II to the fourth assessment report of the Intergovernmental Panel on Climate Change (IPCC) (976 pp.). Cambridge and New York: Cambridge University Press.

MoEF. (2008). *Bangladesh climate change strategy and action plan 2008.* Dhaka: Ministry of Environment and Forests (MoEF), People's Republic of the Government of Bangladesh (GoB).

Index

Note: Page numbers in *italics* indicate figures and photos; **bold** page numbers indicate tables.

9780367584696